Sewer Networks and Processes within Urban Water Systems

WATER AND ENVIRONMENTAL MANAGEMENT SERIES

Sewer Networks and Processes within Urban Water Systems

Selected proceedings of the 18th European and 1st Asian Junior Scientists Workshops organised by the Sewer Systems and Processes Working Group

Edited by J.-L. Bertrand-Krajewski, M. Almeida, J. Matos and S. Abdul-Talib

IWA Publishing

Published by IWA Publishing, Alliance House, 12 Caxton Street, London SW1H 0QS, UK
Telephone: +44 (0) 20 7654 5500; Fax: +44 (0) 20 7654 5555; Email: publications@iwap.co.uk
Web: www.iwapublishing.com

First published 2004

British Library Cataloguing in Publication Data
A CIP catalogue record for this book is available from the British Library

Library of Congress Cataloging-in-Publication Data
A catalog record for this book is available from the Library of Congress

Disclaimer

ISBN: 1 84339 506 1
ISSN: 1476-1785

Contents

Foreword

In September 2002, the SS&PWG (Sewer Systems and Processes Working Group) of the IWA/IAHR Joint Committee on Urban Drainage decided to organize two Junior Scientists Workshops in Europe and Asia. Both workshops were conceived according to the now traditional rules of Juniors Scientists Workshops, as described by Schilling (1999).

The 18th European Junior Scientists Workshop (EJSW) has been held in Almograve, Portugal, on 8-11 November 2003, on the theme “Sewer processes and networks”. There were 24 junior participants from 11 countries and 4 seniors to facilitate the organisation. The main topics were integrated modelling of urban water systems, measurements and modelling of combined sewer overflows, modelling of pollutant loads emitted by sewer systems during dry and wet weather, calibration of models, bed-load transport, sewer pipe roughness, advection in sewers, anoxic processes and oxygen transfer in sewers, infiltration and exfiltration.

The 1st Asian Junior Scientists Workshop (AJSW) has been held in Malacca, Malaysia, on 7-10 February 2004, on “Sewer network processes and urban drainage”, a slightly wider theme to account for regional research activities. There were 19 junior participants from 5 countries and 2 seniors. The main topics were runoff source control, rainfall runoff modelling, pollutant loads in sewers, bed-load transport in sewers and rivers, anoxic transformations in sewers, ventilation and oxygen uptake rates in sewers, deposits and exfiltration in sewers. The AJSW was the first junior workshop with this format in Asia, and interesting perspectives exist to organise future similar events in this region.

In total, 37 papers have been presented in both workshops, which are now accessible to all participants from Asia and Europe through a restricted access website. This will facilitate mutual information and share of knowledge between participants from these two regions. There are similarities in research topics, but with significant adaptations / differences due to climate conditions (e.g. humid tropical context in Asia compared to more moderate European climate), to urbanisation (e.g. very dense urbanisation in Japan compared to Europe) and to sewer systems (e.g. open storm sewers for monsoon in Malaysia). There are also differences in social behaviour during the workshops, as relations between juniors and seniors are different in Asian and European cultures. As a consequence, there is still much to be learnt by all junior and senior participants through more frequent contact and exposure to different cultural, social and scientific practices.

Among the 37 full papers, 16 papers have been selected by independent reviewers from the SS&PWG for publication in this issue of the WEM book series. They reflect rather well the variety of topics presented during both workshops, and offer an opportunity to the junior authors to have a wider audience for their work.

In both workshops, the junior participants themselves chaired sessions, played the role of advocatus diaboli, and prepared reports for all sessions to identify key scientific elements and remaining questions. This intense involvement in the workshop organisation and management is full part of the Junior Scientists Workshops. In order to help the Juniors in these tasks, the seniors provided some guidelines which were appreciated by all participants. As these guidelines may also be interesting for a wider audience, they are reproduced as an appendix to this issue.

Last but not least, these workshops would not have been a success without the significant participation of many SS&PWG Committee members (for more information about SS&PWG activities, see at the website http://www.sspwg.civil.auc.dk/) and of some external senior colleagues, who reviewed the abstracts and the full papers with strong motivation and seriousness. Special acknowledgements to Aminuddin Ab. Ghani (Malaysia), Richard Ashley (UK), Sylvie Barraud (France), Guang-Hao Chen (Hong Kong), Bernard Chocat (France), François Clemens (The Netherlands), Ana Deletic (Australia), Thomas Ertl (Austria), Kapil Gupta (India), Thorkild Hvitved-Jacobsen (Denmark), James Joyce (USA), Ole Mark (Denmark), Jiri Marsalek (Canada), Erik Ristenpart (Germany), and Peter Vanrolleghem (Belgium). Special thanks also to Emilie Baer: she played a key role in relations with the authors and in the final editing work to prepare these proceedings.

Jean-Luc Bertrand-Krajewski, France (Editor)
Maria Almeida, Portugal
José Matos, Portugal
Suhaimi Abdul-Talib, Malaysia

Lyon, France, 22 July 2004

References

Schilling W. (1999). The European Junior Scientist Workshops - a flying start for doctoral students. *Water Science and Technology*, **41**(2), 93-98.

Part One: Selected Papers

Modelling the Integrated Urban Drainage Systems

G. Mannina*, G. Viviani* and G. Freni*

Dipartimento di Ingegneria Idraulica ed Applicazioni Ambientali, Università di Palermo, Viale delle Scienze, Palermo 90128, ITALIA
E-mail: mannina@idra.unipa.it, gviv@idra.unipa.it, freni@idra.unipa.it

Abstract In the last years, several researches were developed to analyse the problems linked with the polluting impact on water bodies because of the pollution produced by rainfall in the urban environment. In order to obtain a good description of the problem, it is important to analyse both quantity and quality aspects connected with all the transformation phases that characterise the water cycle in the urban drainage systems. In the paper, an integrated model has been developed which is able to estimate both the interactions between sewers system, WWTP, CSO and receiving water body and the modifications, in terms of quality, that urban storm water causes inside the receiving water body. Such a model is made up of three sub-models: the rainfall-runoff and flow propagation sub-model in sewer system, which is able to evaluate the qualitative-quantitative characteristics of storm water; the WWTP sub-model, which is representative of the treatment processes; the receiving water body sub-model, that simulate the pollution transformations inside the water body. The proposed integrated model has been applied in order to evaluate the impact of urban developments on the Oreto river near Palermo city (Italy).

Keywords Integrated model, urban drainage, sewer system, WWTP, receiving water body.

INTRODUCTION

Integrated modelling can be defined as modelling able to simulate the interaction between two or more physical component of an urban drainage system, i.e. sewer system (SS), wastewater treatment plant (WWTP) and receiving water body (RWB). Integrated modelling development is basically connected with the attempt of simulating polluting impacts on receiving water body and, at the same time, the possible causes that generates them. This approach is usually aimed to a better management of all the components of the sewer system and to a better knowledge of the transformations of sewage in the different parts of the system (Rauch et al., 2002).

The need for analysing all the aspects of urban drainage systems is also represented in the EU Water Framework Directive 60/2000, that basically proposes a water-quality orientated view on the whole system and requires new ways of assessing their performance. The key to the required consideration is the change from an emission based approach (Emission Standard) to an ambient water-quality based approach (Stream Standard) (Chave, 2001).

Initially, integrated models have been limited to sewer system and WWTP simulation, neglecting receiving water body (Durchschlag, 1990); only in a second time, the integrated approach has been extended to the whole system; a comprehensive description of this kind of models has been provided by EU COST-682 working group (Vanrolleghem et al, 1999).

The most part of developed models couples separated approaches for simulating different system components using the output of each of them as input for the downstream element of the integrated system. As an example, the Integrated Catchment Simulator (ICS), basically links existing commercial models such as MOUSE for drainage systems (Danish Hydraulic Institute, 2003), STOAT for WWTP simulation and MIKE11 for rivers (Tomicic et al., 2000). Other examples of integrated models are WEST simulator for biochemical processes analysis and SIMBA that was developed in MATLAB/SIMULINK environment (Coen et al., 1997; Alex et al., 1999). The basic limitation of commercial models is connected to the unavailability of model source code, not allowing the final user for modifying and further developing model options in order to fit specific simulation needs.

Another problem frequently linked with commercial models regards the high parameterisation that requires long and expensive data acquisition activities (Vanrolleghem et al., 1999). For by-passing this kind of problem, Schutze et al. (1999) proposed the use of "semi-hypothetic" case studies for analysing integrated modelling potentialities. In details, the Authors simulated the different parts of the system (SS, WWTP, RWB) using data coming from real and well documented case studies not really linked together.

In the paper, an integrated model has been developed which is able to estimate both the interactions between the different systems and the modifications, in terms of quality, that urban storm water causes inside the receiving water body. The model simulate all urban drainage system taking into account the most important phenomena that regulate its behaviour with the objective to reduce the number of the parameters making its use, in real cases, easier. The model is able to simulate both synthetic and historical events.

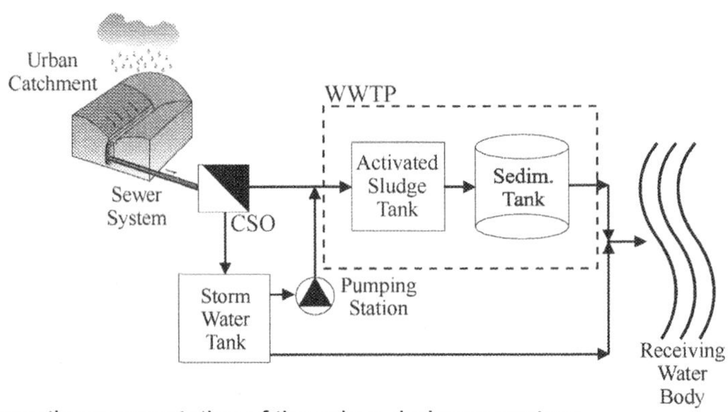

Figure 1: Schematic representation of the urban drainage system.

MODEL DESCRIPTION

The model is made up mainly of three sub-models (Figure 1):

a) the rainfall-runoff and flow propagation sub-model, which is able to evaluate the qualitative-quantitative characteristics of storm water in the sewer system; a CSO device was introduced at the outlet of the sewer system in order to contain the flow excursion in the WWTP during the storm event. To contain the CSOs and to achieve a RWB protection, a storm water tank (SWT) has been introduced.
b) the WWTP sub-model, which is representative of the treatment processes; the WWTP units considered in the sub-model consist of the activated sludge tank and

the secondary sedimentation tank, that are the main units influenced by quali-quantitative characteristic variations of the storm sewage flow.

c) the receiving water body sub-model, that simulates the pollutants transformations inside the water body; the adopted algorithms are able to analysis of ephemeral natural stream that is typical of southern Europe.

Each sub-model has been set up according with the relative theory inherent the different part of the model to achieve a good simulation of the different processes take place in the whole system. All part of the system are computed simultaneously. The system compartments (SWT and WWTP) were previously designed in steady dry conditions. The main used algorithms are reported in Table 1. Table 2 summarised the algorithms parameters and variables, their physical meaning and the proposed values used for dynamic simulation.

Sewer system

This sub-model is able to simulate the main phenomena that take place both in the catchement and in the sewer network during a storm event. It is divided into two connected modules: a hydrological and hydraulic module, which calculates the hydrographs at the inlet and at the outlet of the sewer system, and a solid transport module, which calculates the pollutographs at the outlet for different pollutants (TSS, BOD and COD). The hydrological and hydraulic module starts to evaluate the net rainfall, from the hyetograph measured, by a loss function (initial and continuous). From the net rainfall, the model simulates the rainfall-runoff process and the flow propagation with a cascade of two reservoirs in series (eq.1 in Tab. 1).

The solid transfer module reproduces the accumulation and propagation solids in the catchment and in the sewer network. The main phenomena simulated are: build-up and wash-off of pollutants from catchment surfaces, sedimentation and re-suspension of pollutants in sewers (Bertrand-Krajewski et al., 1993). To simulate the build-up on the catchment surfaces an exponential function (eq. 2) was adopted (Alley and Smith, 1981). The solid wash-off caused by overland flow during a storm event was simulated with the formulation in eq. 3 proposed by Jewell and Adrian (1978).

The solids deposits in the sewers during dry weather has been evaluated by adopting an exponential law (eq. 4); this phenomenon has been emphasized in the model because of its contribution to the storm water pollution. Two classes of particles have been considered: fines particles (d_{50}=50 μm, specific gravity s=1,6) and the coarse particles (d_{50}=500 μm, specific gravity s=2,0). The first classes are mainly transported in suspended load instead the second one are transported in bed load or suspended load in relation of the characteristics of the flow. In order to have a realistic and correct approach a particular care has been taken about sediments transformation in sewers, considering their cohesive-like behaviour linked to organic substances and to the physical-chemical changes during the sewer transport (Crabtree, 1989; Ristenpart, 1995). In particular, the transport equation proposed by Parchure and Metha (1985) couplet to the bed sediment structures hypothesised by Skipworth et al. (1999) to simulate the sediment erosion rate was considered (eq. 5). The pollutographs at the outlet of the sewer system have been evaluated by hypothesising the complex catchement sewer network as a reservoir and by considering an adapted version of Wiuff's model (Berthrand-Krajewski, 1993).

Finally, the WWTP inflow has been computed taking into account the presence of CSO device: its behaviour has been simulated by the eq. 6, where CSO efficiency has been taken into account by the k_{CSO} coefficient. The SWT inflow and outflows quali-quantitative characteristics has been considered by the eq. 7-8.

Wastewater treatment plant

Substrate and micro-organisms concentration in the activated sludge tank have been calculated with mass balances based on Monod's theory (eq 9-10). The sedimentation tank performance has been simulated using the solid flux theory according to the methodology proposed by Takács et al. (1991). In particular, the solids concentration profile has been obtained by dividing the settler into 50 horizontal layers of constant thickness. Within each layer the concentration is assumed to be constant and the dynamic update is performed by imposing a mass balance for each layer. Three sets of equations have been adopted for the layers depending on their depth: the first set (eq. 11) has been used for the upper region of the tank (the clarification region), the second one (eq. 12) for the lower layer (the thickening zone) and the third for the feed layer (eq. 13). The settling velocity function (eq. 14) proposed by Takács et al. (1991) has been adopted.

Receiving water body

This sub-model simulates the processes that take place in the river. It is divided into quantity and quality module. The first module allows to determinate the flow distribution in the river instead the second module simulates the distribution of a generic pollution along the river. The exemplified form of the De Saint Venant equation (cinematic wave) for the quantity module and the dispersion advection equation for the quality module were adopted (eq.15-16). The dispersion coefficient E has been evaluated by using the expression suggested by Elder (eq. 17), where only the vertical velocity gradient is considered important in stream flow (Brown and Barnwell, 1987).

APPLICATION OF THE MODEL

The model has been applied to a semi-hypothetical case study; this approach was developed to cope with the lack of date and to verify, in this first phase, the performances of the model (Schutze et al., 1999). The model has been tested by using the databases of Fossolo (Bologna) catchment, for the sewer system, and Oreto river (Palermo) for the receiving water body. Fossolo experimental catchment collects water coming from a residential area on the outskirts of Bologna (Artina et al., 1997). The drained area measures 40.71 ha, with an impermeable percentage of 75%. The catchment has an average slope of 0.3%. The number of inhabitants is about 10.000. The drainage network length is about 5 km and it ends in a polycentric section pipe 144 cm high and 180 cm wide. Discharge has been estimated by an ultrasonic probe. A refrigerated automatic sampler with 24 bottles has been used for sewage quality analysis. 12 events have been measured.

Oreto catchment area is about 112 km^2. The river has an ephemeral natural stream flow with alternation of intense short rainfall and long dry periods. The upstream reach is still natural instead the downstream part has been urbanized during the last 50th years. The upstream reach has got a strong auto-depuration capacity due to its turbulent hydraulic regime. Twelve cross section have been selected in order to estimate the whole stream behaviour in terms of pollutants and dissolved oxygen concentration.

SEWER SYSTEM

$$Q_i = \int_0^t Q_j(\tau)\frac{1}{k_i}e^{-\frac{t-\tau}{k_i}}d\tau \quad i,j=1,2 \quad Q_j = \begin{cases} Pn & if \quad j=1 \\ Q_1 & if \quad j=2 \end{cases} \tag{1}$$

$$Ma = \left(\frac{Accu}{Disp}\right)\cdot A\cdot Peim\cdot\left(1-e^{-\left(\frac{Disp\cdot ts}{24}\right)}\right)+Mr\cdot e^{-\left(\frac{Disp\cdot ts}{24}\right)} \tag{2}$$

$$M_{ent} = M_{at}\cdot(1-e^{-(Arra\cdot Pn^{Wh}\cdot\Delta t)}) \tag{3}$$

$$M_{dep} = CS + (M_0 - CS)\times e^{-K_{dep}t_s} \tag{4}$$

$$M_{ero} = M\left(\frac{\tau_b-\tau_c}{\tau_c}\right) \tag{5}$$

$$\begin{cases} Q_{imp} = Q_{\max,p,\lim} + (Q_{\max,p,\lim} - Q_{\max,tP})\cdot(1-e^{-(k_{CSO}\cdot Q_2)}) & Q_2 > Q_{\max,tp} \\ Q_{imp} = Q_2 & Q_2 \le Q_{\max,tp} \end{cases} \tag{6}$$

$$\frac{dW}{dt} = Q_{s,in} - Q_{s,out} \tag{7}$$

$$\frac{dC_iW}{dt} = M_{S,in,i} - M_{S,out,i} \tag{8}$$

WASTEWATER TREATMENT PLANT

$$V\frac{dS}{dt} = Q_{imp}\cdot S_F - \nu\cdot X\cdot V - Q_{imp}\cdot S_E \tag{9}$$

$$V\frac{dX}{dt} = Q_{imp}\cdot X_F + (\mu - k_d)\cdot X\cdot V + Q_R\cdot X_R - (Q_{imp}+Q_R)\cdot X \tag{10}$$

$$\begin{cases} \dfrac{dXi}{dt} = \dfrac{J_{up,i+1}+J_{clar,i-1}-J_{up,i}-J_{clar,i}}{z_i} & J_{up,i} = \nu_{up}\cdot X_i \quad \nu_{up} = \dfrac{Q_e}{A_S} \\ \begin{cases} J_{clar,i} = J_{s,i} & if \quad X_{i+1}\le X_t \\ J_{clar,i} = \min(J_{s,i},J_{s,i+1}) & if \quad X_{i+1} > X_t \end{cases} \end{cases} \tag{11}$$

$$\frac{dXi}{dt} = \frac{v_{dn}(X_{i-1}-X_i)+\min(J_{s,i},J_{s,i-1})-\min(J_{s,i},J_{s,i+1})}{z_i} \quad J_s(X_i)=v_s(X_i)\cdot X_i;\ v_{dn}=\frac{Q_u}{A_S} \tag{12}$$

$$\frac{dXi}{dt} = \frac{\dfrac{Q_fX_f}{A_S}+J_{clar,i+1}-(v_{up}+v_{dn})\cdot X_i-\min(J_{S,i},J_{S,i-1})}{z_i} \tag{13}$$

$$v_s = \max(0,\min(v_0',v_0(e^{-r_h(X_i-X_{\min})}-e^{-r_p(X_i-X_{\min})}))) \quad where \quad X_{\min}=f_{ns}\cdot X_f \tag{14}$$

RECEIVING BODY

$$\frac{\partial DO}{\partial t}A_X = \frac{\partial\left(A_X\cdot E\dfrac{\partial DO}{\partial x}\right)}{\partial x} - \frac{\partial(A_X\cdot U\cdot DO)}{\partial x} + A_X[k_a(DO_s-DO)-(k_d+k_S)BOD] \tag{15}$$

$$\frac{\partial BOD}{\partial t}A_X = \frac{\partial\left(A_X\cdot E\dfrac{\partial BOD}{\partial x}\right)}{\partial x} - \frac{\partial(A_X\cdot U\cdot BOD)}{\partial x} + A_X[-(k_d+k_S)BOD] \tag{16}$$

$$E = K_{disp}\cdot d\cdot u^* \tag{17}$$

Table 1. Model algorithms.

VARIABLES AND PARAMETERS			M.U.	VALUE
SEWER SYSTEM	Q_n	Average dry flow rate	m^3/h	49
	K_1- K_2	storage constants	min	4.2-5.4
	Ma	Total mass accumulated on the catchment surface	kg	
	t_s	Antecedent dry weather period	h	
	Accu	Build-up rate	kg/(ha*d)	8
	Disp	Decay rate	d^{-1}	0.07
	A	Catchment surface	ha	40.71
	Peim	Percentage of impervious area	-	0.75
	Mr	Residual mass after last event	kg	
	M_{ent}	Mass entering in sewer system in time step Dt	kg	
	M_{at}	Mass on the catchment at the time t	kg	
	Arra	Wash-off coefficient	mm^{-1}	0.18
	Wh	Wash-off factor	-	0.636
	Δt	Time step	sec	60
	Pn	Net rainfall intensity	mm/h	
	M_{dep}	Mass of deposits	kg	
	M_{ero}	Eroded mass in sewer	kg	
	CS	Storage capacity of the sewer system	kg	1580
	M_o	Mass residual in sewer after the end of the last event	kg	
	K_{dep}	Deposition coefficient	h^{-1}	0.0099
	τ_b	Applied bed shear stress	N/m^2	
	τ_c	Critical bed shear stress at point of erosion	N/m^2	4
	M	Erosion rate when $\tau_b=2\tau_c$	kg	7.68
	Q_{imp}	Influent flow rate to WWTP	m^3/h	
	Q_i	Flow rates in the sewer sub-model	m^3/h	
	Q_{maxtp}	Theoretical maximum WWTP flow rate ($3Q_n$)	m^3/h	147
	$Q_{max,p,lim}$	Maximum WWTP flow rate ($4Q_n$)	m^3/h	196
	k_{CSO}	CSO coefficient	h/m^3	0.8
	W	Cumulative volume stored in the SWT	m^3	
	$Q_{S,in}$	Influent SWT flow rate	m^3/h	
	$Q_{S,ou}$	Effluent SWT flow rate	m^3/h	
	C_i	Pollutant concentration	kg/m^3	
	$M_{S,in,i}$	Pollutant inlet SWT mass rate	kg	
	$M_{S,out,i}$	Pollutant outlet SWT mass rate	kg	
	V	Activated sludge tank volume	m^3	1179
	S	Biodegradable substrate concentration	kg/m^3	
	S_F	Effluent biodegradable substrate concentration	kg/m^3	0.4 *
	S_E	Effluent biodegradable substrate concentration	kg/m^3	0.02 *
	v	Substrate removal coefficient	h^{-1}	0.0274
	X	Active heterotrophic biomass concentration in the tank	kg/m^3	4 *
	X_F	Active heterotrophic biomass concentration in the influent	kg/m^3	
	Xr	Concentration in return flow rate	kg/m^3	10
	μ	Yield coefficient for the heterotrophic biomass	h^{-1}	0.0137
	k_d	Decay velocity of heterotrophic organisms	h^{-1}	0.00096

(cont.)

WASTEWATER TREATMENT PLANT	X_i	Suspended solid concentration of the layer "i"	kg/m^3	
	X_t	Suspended solids threshold concentration	kg/m^3	5
	J_{up}	Upward flux of suspended solids due to upward bulk flow	$kgTSS/m^2*h$	
	J_{clar}	Special flux function for the clarification zone of the settler	$kgTSS/m^2*h$	
	J_s	Solid flux due to gravity settling	$kgTSS/m^2*h$	
	v_{up}	Upward bulk fluid velocity	m/h	
	v_{dn}	Downward bulk fluid velocity	m/h	
	Q_e	Settler effluent flow rate	m^3/h	
	Q_R	Return flow rate	m^3/h	56.66
	Q_u	Settler underflow rate	m^3/h	
	A_S	Settler surface area	m^2	115
	H	Settler depth	m	4.2
	FS_{max}	Maximum solids loading rate	$kgSS/m^2*h$	4.9 *
	F/M	Food to microorganism ratio	kgBOD/kgSS*d	0.1 *
	C_i	Surface load on the settler	m/h	1.1 *
	T_r	Hydraulic retention time	h	3 .8 *
	z_i	Height of the layer "i"	m	0.08
	v_S	Settling velocity	m/h	
	v_0'	Maximum (practical) settling velocity	m/h	4.17
	v_0	Maximum (theoretical) settling velocity	m/h	6.04
	X_{min}	Suspended solids minimum concentration	g/l	0.01
	r_h	Settler model parameter for hindered settling	l/gr	0.42
	r_p	Settler model parameter for low solids concentration	l/gr	5
	X_{in}	Feed solid suspended concentration to the settler tank	kg/m^3	4 *
	f_{ns}	Non settleable fraction of X_f		
	r_c	Flow rate ratio	-	0.7
RECEIVING BODY	E	Longitudinal dispersion coefficient	m^2/s	
	d	Mean depth of the stream	ft	
	u*	Average shear velocity	m/s	
	K_{disp}	Dispersion coefficient	-	5.93
	DO	Dissolved oxygen concentration	mg/l	
	DO_S	Dissolved oxygen saturation concentration	mg/l	10
	A_X	Cross sectional area	m^2	
	U	Average velocity	m/s	
	k_a	Reareation coefficient	s^{-1}	0.004255
	k_d	Deoxigenation coefficient	s^{-1}	0.0005
	k_s	Sediment oxygen demand	s^{-1}	0.000037

Table 2. Model variables and parameters (* indicates the values adopted in the design procedure).

RESULTS

The model has been applied using the 12 events recorded in Fossolo catchment. Fig 2 shows the hydrograph and pollutograph in sewer for the event recorded on 25/4/94. The pollutograph is characterized by a first flush effect. Fig. 3 shows the CSO pollutograph during the storm event. At the first part of the event the pollution load rises rapidly; this causes the fill of the retention tank and the discharge in the RWB. In Fig. 4 the settler behaviour is shown. Fig.5 shows flows and dissolved oxygen concentrations along the water body, in different sections of the river.

The model allows for evaluating the variations of the WWTP performances during a storm event and the dissolved oxygen reduction in the RWB due to the discharge. In particular, the increase of flow rates causes a reduction in treatment efficiency both regarding the activated sludge tank and the sedimentation tank. This fact can cause the overflow of TSS and then the consequent discharging in the receiving water body. Moreover, the loss of the biomass from the system causes a general reduction in the treatment process efficiency worsening the quality conditions of the receiving water body also in the dry period subsequent the event. Indeed normal working conditions can't take place immediately after the end of the storm event but it involves usually a time period, for the re-establishment of the biomass in the activated tank reactor.

The consequent dissolved oxygen reduction in the RWB is much more emphasized in the upstream sections, near the WWTP and SWT outlets. This aspect could cause the water fauna dead and the consequent decay of the river environmental quality characteristics.

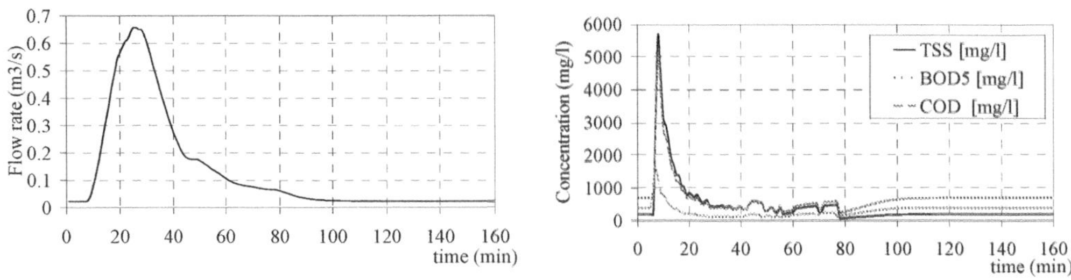

Figure 2: Hydrograph and pollutographs in sewer.

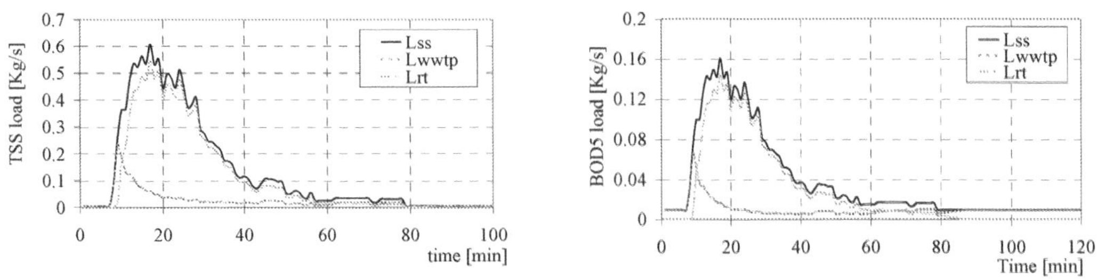

Figure 3: Pollutographs in the SS outflow, in the CSO and in the WWTP inflow.

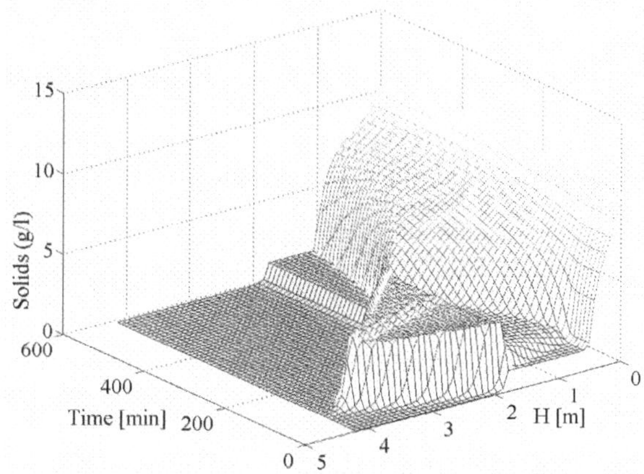

Figure 4: TSS profiles in sedimentation tank.

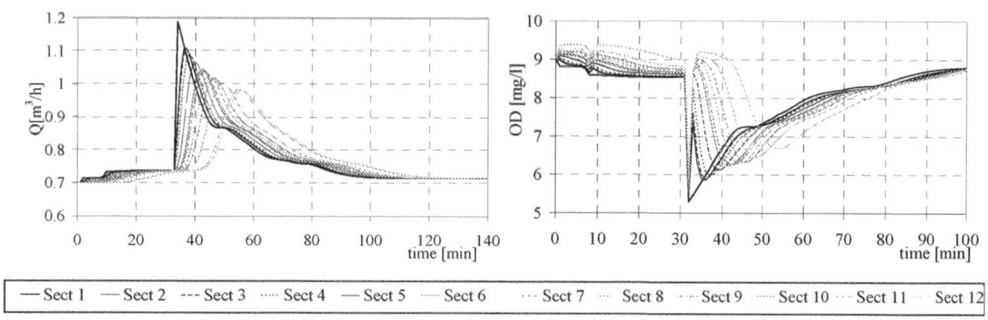

Figure 5: Hydrographs and temporal variation of DO in selected cross-sections along Oreto river.

CONCLUSIONS

An integrated model has been developed to analyse the impact on RWB quality coming from urban sewage discharge. In this first phase of the research the model has been applied to an hypothetical site to test the accuracy during different storm events. The model is able to estimate the effect of the introduction of mitigation and control measures to reduce the CSOs. In order to preserve the water body, the temporal accumulation of the first part of the event in a SWT and its treatment at the end of the storm event has been considered.

The model adopts simple but accurate algorithms to reproduce main phenomena that characterize the integrated system component behaviour. This approach allows to overcame problem of lack of data due to the several processes to simulate. Further development of the research will be given to the application of a real pilot case.

REFERENCES

Alex, J., Risholt, L.P., Schilling, W. (1999). Integrated modeling system for simulation and optimization of wastewater systems. *Proceedings of 8th International Conference on Urban Storm Drainage, Sydney, Australia*, 1553-1561

Alley, W.M., Smith, P.E. (1981). Estimation of accumulation parameters for urban runoff quality modelling. *Wat. Reso. Res.*, **17**, 1657-1664.

Artina, S., Maglionico, M., Marinelli, A., Raffaelli, G., Anzalone, C., Lanzarini, S., Guzzinati, E. (1997). Le misure di qualità nel bacino urbano Fossolo. L'acqua 2, 17-25.

Bertrand-Krajewski, J.L. (1993). Modélisation du transport solide en réseau d'assainisse-ment unitaire: le modèle Hypocras. *La Houille Blanche* **4**.

Bertrand-Krajewski, J.L., Briat P., Scrivener O. (1993). Sewer sediment production and transport modelling: A literature review. *J. Hydr. Research* **31**(4), 435-460.

Brown, L.C., Barnwell, T.O. (1987). The enhanced stream water quality models QUAL2E and QUAL2E-UNCAS: documentation and user manual, EPA.

Chave, P. (2001). *The EU Water Framework Directive: an introduction.* IWA Publishing, London.

Coen F., Vanderhaegen B., Vanrolleghem P.A. and Van Meenen P. (1997). Improved design and control of industrial and municipal nutrient removal plants using dynamic models. *Wat. Sci. Tech.*, **35**(10), 53-61.

Crabtree, R. W. (1989). Sediments in sewers. J. Inst. Water and Environmental Management, **3**, 569-578.

Danish Hydraulic Institute DHI (2003). MOUSE TRAP User Guide. *www.dhisoftware.com/mouse/Download/Documentation.htm*

Durchschlag, A. (1990). Long-term simulation of pollutant loads in treatment plant effluents and combined sewer overflows. *Wat. Sci. and Tech.*, **22**(10-11), 69-76.

Jewell, T.K., Adrian, D.D., Asce, M. (1978). SWMM storm water pollutant washoff function. *Journal of the Environmental Engineering Division*,104, 5.

Parchure, T.M., Mehta, A.J. (1985). Erosion of soft cohesive sediment deposits. *J. Hydr. Eng., ASCE*, **111**, No. 10.

Rauch, W., Bertrand-Krajewski, J.-L., Krebs, P., Mark, O., Schilling, W., Schutze, M., Vanrolleghem, P. A. (2002). Deterministic modelling of integrated urban drainage system. *Wat. Sci. and Tech.*, **45**,(3), 81-93

Ristenpart, E. (1995). Sediment properties and their changes in a sewer. *Wat. Sci. Tech.*, **31** (7).

Schutze, M., Butler, D., Brucee Beck, M. (1999). Optimisation of control strategies for the urban wastewater system. An integrated approach. *Wat. Sci. and Tech.*, **39**(9), 209-216.

Skipworth, P.J., Tait, S.J., Saul, A.J. (1999). Erosion of beds in sewers: Model development. *J. Hydr. Division, ASCE*, **125**, No. 6.

Takács, I., Patry, G.G., Nolasco, D. (1991). A Dynamic Model of the Clarification-Thickening Process. *Wat. Res.*, **25**(10), 1263-1271.

Tomicic B., Clifforde I.T., Mark O., Yde L., Dudley J. (1999). Integrated Catchment Simulator (ICS) – Development and applications. *3rd DHI Software Conference (www.dhi.dk/softcon/index.htm)*

Vanrolleghem P., Schilling W., Rauch W., Krebs P., Aalderink H. (1999). Setting up measuring campaigns for integrated wastewater modelling. *Wat. Sci. and Tech.*, **39**(4), 257-268.

Model connectors for integrated simulations of urban wastewater systems

L. Benedetti, J. Meirlaen and P.A. Vanrolleghem

BIOMATH, Ghent University, Coupure Links 653, B-9000 Gent (Belgium) tel. +32 9 264 6196, fax. +32 9 264 6220, e-mail lorenzo.benedetti@biomath.ugent.be

Abstract In order to perform simulations of the integrated wastewater system, state variables of sub-system models (sewer, activated sludge, river) must be translated from one model to the other. A connector between ASM1 (Henze *et al.* 2000) and RWQM1 (Shanahan *et al.* 2001) is described in this paper to introduce the general approach developed to overcome this necessity. Inherent features of this connector are its closed elemental mass balances. The COD fractions of ASM1 have been split over the COD fractions of RWQM1, while balance terms were used to close elemental balances. Also the different environmental conditions in the systems (activated sludge, sewer and river) have been taken into account. An evaluation of the influence of several connector and model parameters on connector balance terms and outputs has been performed. Some results of a case study concerning the river Lambro (Italy) are described to show an application of such model connection approach.

Keywords ASM1, integrated modelling, mass balances, model connectors, RWQM1, urban wastewater system.

1 Introduction

In recent years the possibility of performing integrated simulations has largely improved due to the enhanced models described in literature and the continuous development of simulation software together with the increasing computational power at the desktop. Recently research on the integrated urban drainage system has received ever-increasing attention, especially after the approval by the European Union of the EU Water Framework Directive.

The integration of the different components of the urban drainage system in a single model allows the better understanding of (1) the working of the system as a whole, and (2) the mutual interaction between its components. This is clearly essential for the evaluation of the system performance as well as for the detection of the system's weak points, in the context of the development of environmentally and economically sustainable planning and management practices. It is also a necessary step for the implementation of Real Time Control (RTC) strategies for the urban drainage system (Meirlaen *et al.* 2001).

One of the main problems when developing an integrated model is the incompatibility between state variables, processes and parameters used in the different sub-models. An example of the difference between state variables is given by the treatment plant model and the river model, the former being typically based on COD and the latter on BOD, like QUAL2E (Brown and Barnwell 1987). (Maryns and Bauwens 1997) tried to avoid the problem by using the ASM1 model in riverine conditions but this approach did not give satisfactory results. To tackle this problem more fundamentally, the IWA task group on river water quality has developed a new COD based model, RWQM1 (Reichert *et al.* 2001). The

 Sewer Networks and Processes within Urban Water Systems edited by J.-L. Bertrand-Krajewski, M. Almeida, J. Matos & S. Abdul-Talib. ISBN: 1 84339 506 1.

states of this model are more like the ASM states, but some differences still remain. This is due to the fact that the full RWQM1 model has to describe more components than the ASM1 model (e.g. algae growth).

The environmental conditions in the sub-systems being rather different, the main processes will also differ due to the fact that the same group of organisms behaves differently depending on the environment. For instance, nitrifying bacteria in an activated sludge system are confronted with a high competition for oxygen due to the presence of heterotrophic organisms in high concentrations, while in the river system this competition is not that strong.

Therefore, when using two models with different state variables, a connector needs to be developed in order to 'translate' the state variables of the original model to the state variables of the destination model in a consistent way. This connector needs to contain all available knowledge about the different states in the two models, different environmental conditions in origin and destination, and requires closed elemental balances; in most cases the considered elements are C, N, P, O and H. This elemental balancing feature is one of the basic principles in all types of models, and it is hence important to adhere to these principles. This can also be seen from the evolution of activated sludge models, where ASM2 and ASM3 models (Henze *et al.* 2000) have closed elemental mass balances as well. Unlike a few years ago, when all in-sewer process models were BOD-based (Fronteau *et al.* 1997), recent models for this sub-system follow the same tendency and are COD-based as well, which makes integration and connection of models for the urban wastewater system easier (Huisman *et al.* 2003; Mourato *et al.* 2003).

In this paper, first a description is made of the connector between the state variables of ASM1 (as representative of sewer or WWTP process models) and the state variables of RWQM1. The approach adopted to create this connector can be used for any other coupling of similar models. Then some considerations follow on the influence of the state variables elemental compositions and connector parameters on elemental balances and connector output. A case study on integrated modelling of the effects of Combined Sewer Overflow (CSO) and WWTP effluent on the river water quality of the river Lambro (Italy) is presented to illustrate an application of such connector.

2 The model connector

The connector is a list of algebraic equations expressing concentration inputs in the river in terms of concentration outputs from the sewer or WWTP.

RWQM1 uses COD as a measure for organic pollution, which is in contrast with the more traditional river water quality models. This makes integration with the also COD-based ASM models easier. Moreover, RWQM1 has closed mass balances for COD, and closed elemental balances for C, N, P, O and H. Since these are important properties, the goal was to keep these in the new connector.

Due to the modelling approach used in RWQM1, some state variables can be very easily transformed from the activated sludge to the river conditions. Slowly and readily biodegradable substances will probably remain biodegradable when entering the river. One could also suppose that the active biomass coming from the sewer or activated sludge system would remain active biomass in the river. However, this is probably not completely true since the environmental conditions in the sewer or treatment plant and the river are usually very different in terms of temperature, food, light intensity, etc. (as a consequence these three variables may also vary with the period of the year), probably causing inactivity of at least part of the biomass. In the connector, biomass is split into a part that remains active and a part that is transformed into inert and slowly biodegradable organic matter. This fraction can be assumed to be bigger for biomass from sewage than from activated sludge.

Autotrophic biomass is modelled as first step and second step nitrifiers in RWQM1, but only as one group of organisms in the ASM1. In the connector the incoming autotrophic biomass is split into two active groups, first and second step nitrifiers (with respective surviving fractions f_{N1} and f_{N2}), while the remaining (dead) part is split into slowly biodegradable and inert particulate organic matter. Parameters for the fractions are to be found by calibration, since so far no values are available in literature, only the relationship $f_{N1} = 3 \cdot f_{N2}$ (Focht and Verstraete 1977) can be useful as a first approximation. Dead biomass is split in two particulate fractions, one biodegradable and one inert, calculated by means of the parameter f_I, that is likely to be similar to the parameter f_P present in ASM1, since the concept is exactly the same, even if it has to be considered that environmental conditions are different.

The way in which particulate matter and soluble components is transformed when going from the treatment plant to the river is shown in Figure 2.1.

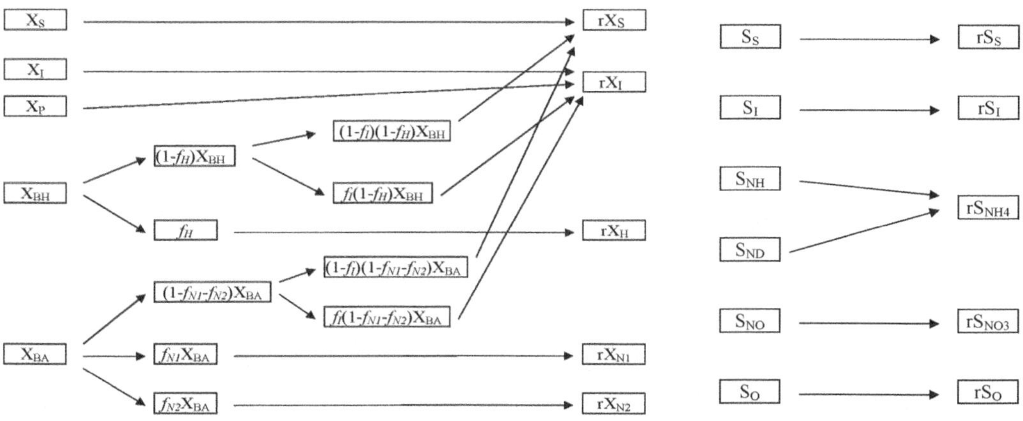

Figure 2.1 Fate of particulate biomass (left) and of soluble components (right) from state variables of ASM1 to RWQM1.

To calculate the output concentrations of the three species of carbonates (CO_2, HCO_3^- and CO_3^{2-}) in the river model, the carbonate equilibrium equations have been implemented, they are a function of incoming alkalinity (state variable in ASM1) and pH, which has been considered as a parameter of the connector.

Soluble organic nitrogen (S_{ND} in ASM1) is added to NH_4 in the output, since the ammonification process is usually fast, and no similar state is present in the river model.

Variables that are used in the RWQM1, but not in the ASM1, like nitrite, particulate phosphate or algae, are set by fixed parameters, which need to be estimated for the system under study.

After finishing all these logical transformations that maintain closed mass balances, elemental balances still need to be closed. This is done using compensation terms, which are used to compensate for a lack or surplus of elements. To this end, the elemental flux calculated in ASM1 state variables should be compared to the elemental flux calculated in RWQM1 state variables. The difference between these two fluxes (either negative or positive) is the compensation term, and needs to be added to the state chosen to serve as a balance term for the specific element. In an ideal set of models and connectors, compensation terms should always be zero. A balance term serves as a sink or source of elements in the organic compounds, e.g. if more nitrogen is present in the organic matter entering the river than is coming from the treatment plant, then the amount of ammonia going into the river is artificially decreased in order to close the nitrogen balance over the connector. The balance terms chosen in the river model are carbonates for carbon, NH_4^+ for nitrogen, HPO_4^{2-} for phosphorus, DO for oxygen and H^+ for hydrogen. To calculate the flux of the five elements coming from ASM1 one needs to fix the composition of the different model components. The nitrogen and phosphorus content were taken according to (Henze *et al.* 2000), while carbon, oxygen and hydrogen content were taken according to (Reichert *et al.* 2001). The following equation illustrates the use of a compensation term for nitrogen, which in the connector is added to the outgoing NH_4^+ balance term.

$$N\_comp = \begin{bmatrix} i_{N\_SS} \cdot S_S + i_{N\_SI} \cdot \mathrm{S_I} + i_{N\_XBH} \cdot \mathrm{X_{BH}} + i_{N\_XBA} \cdot \mathrm{X_{BA}} + i_{N\_XP} \cdot \mathrm{X_P} \\ -i_{N\_rSS} \cdot rS_S - i_{N\_rSI} \cdot \mathrm{rS_I} - i_{N\_rXH} \cdot \mathrm{rX_H} - i_{N\_rXN1} \cdot \mathrm{rX_{N1}} \\ -i_{N\_rXN2} \cdot \mathrm{rX_{N2}} - i_{N\_rXS} \cdot \mathrm{rX_S} - i_{N\_rXI} \cdot \mathrm{rX_I} \end{bmatrix}$$

or

$$N\_comp = \sum_{J}^{ASM1} i_{N\_J} \cdot J - \sum_{J}^{RWQM1} i_{N\_J} \cdot J$$

in which i_{N_J} is the nitrogen content of component J
S_i, X_i are components of the ASM1 model ($gCOD/m^3$)
rS_i, rX_i are components of RWQM1 ($gCOD/m^3$)

3 Results and discussion

The two process models and relative connector have been implemented in the WEST® modelling and simulation platform (Vanhooren *et al.* 2003), which allows to model the integrated urban wastewater system in a single platform, thus avoiding software communication and file transfer problems.

First, a number of simulations have been performed to test the influence of several factors on the connector performance. To represent the behaviour of a combined sewer overflow discharging directly in a river reach, the dry weather influent file for the 'benchmark' (Spanjers *et al.* 1998) has been used as input to the connector between the sewer system and a river stretch. The values of the adjustable parameters in the connector are specified in Table 3.1:

Table 3.1 Values for adjustable connector parameters

parameter	description	value
f_H	fraction of active heterotrophs	0.5
f_{N1}	fraction of active first step nitrifiers	0.15
f_{N2}	fraction of active second step nitrifiers	0.45
f_S	fraction of biomass that becomes inert particulate	0.08
pH		7

In Figure 3.1 the magnitude and evolution of the compensation terms is presented. It can be noticed that only carbon has a positive mass balance, while all other elements have a rather negative balance. This is due to the strong influence of the elemental composition of certain state variables on the elemental balances. A consequence is that, as evident in Figure

3.2, the concentration of some river components coming out of the connector is calculated to be negative, which is, of course, not acceptable.

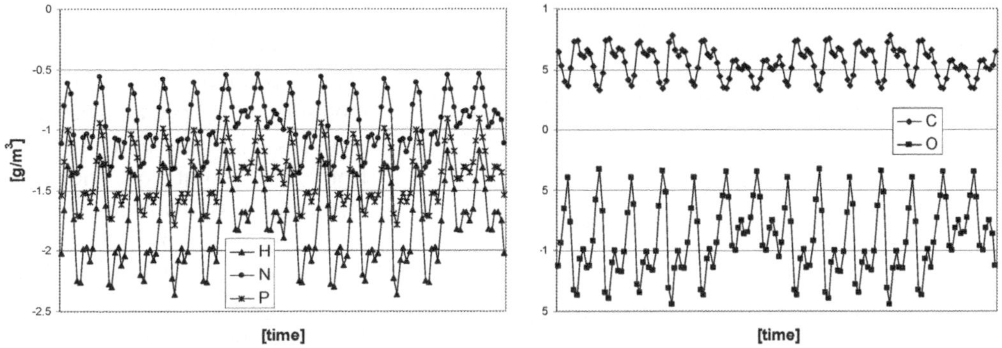

Figure 3.1 Dynamic profiles of compensation terms.

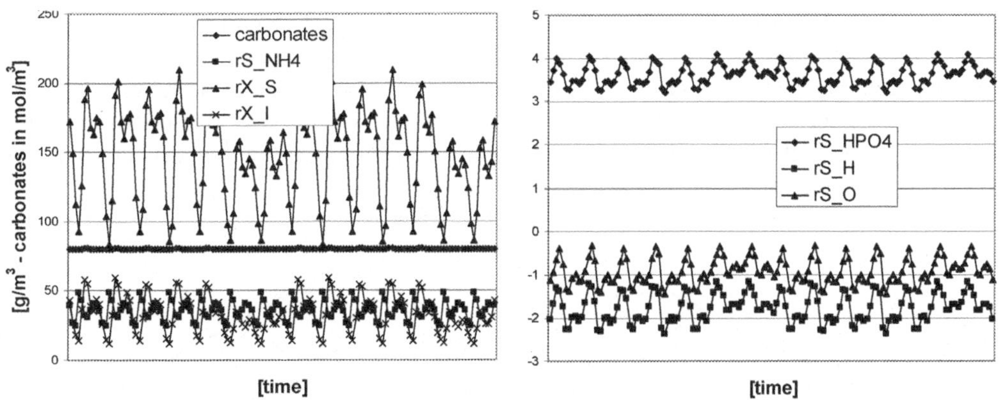

Figure 3.2 Dynamic profiles of relevant connector outputs.

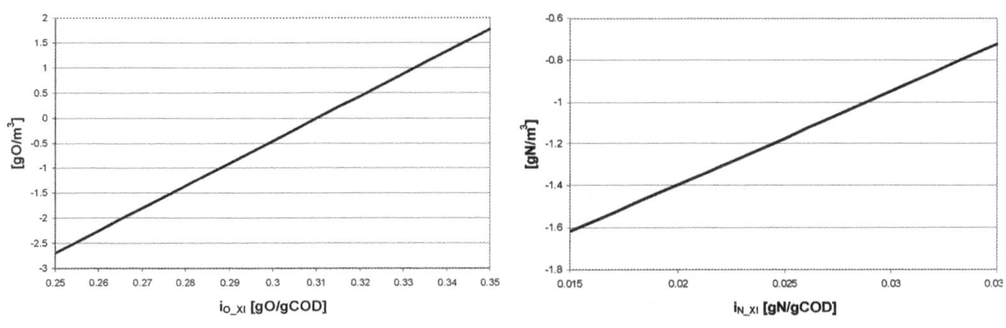

Figure 3.3 Influence of i_{O_XI} and i_{N_XI} on O (left) and N (right) balances.

As an example, the influence of the parameters indicating oxygen and nitrogen contents in inert particulate matter (i_{O_XI} and i_{N_XI}) on oxygen and nitrogen balances has been simulated. Results are shown in Figure 3.3. It is to be noticed that assuming the reference value of i_{N_XI} in ASM2 (0.3 gN/gCOD) or in ASM2d (0.2 gN/gCOD; (Henze *et al.* 2000)) leads to a difference in the nitrogen balance of almost 0.5 gN/m^3. It is evident from the oxygen balance that small changes in this type of parameters have effects of paramount importance for the

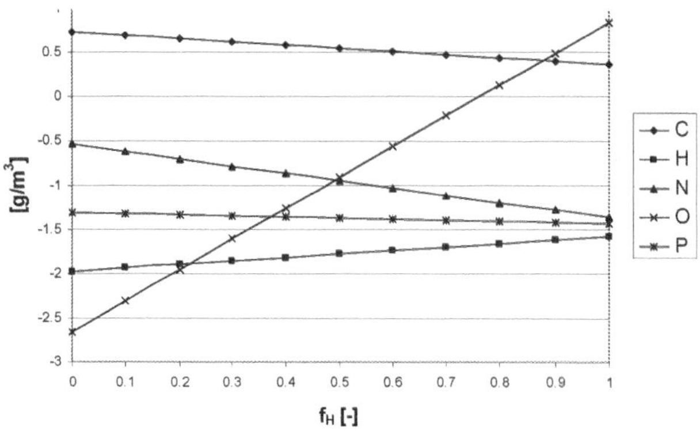

Figure 3.4 Influence of f_H on elemental mass balances.

elemental mass balances. A consequence is that the fractionation in model state variables of measurements of COD, BOD, TN, TP and similar data, coming from routine monitoring or special laboratory analyses, may also be considered from the point of view of elemental mass balances. This is especially true in the case of integrated modelling with conversion of state variables from one model to another.

The most influencing parameter in the connector is probably f_H (fraction of active heterotrophs), the effect of which on elemental balances and connector relevant outputs is shown in Figures 3.4 and 3.5 respectively. Since the oxygen content in the benchmark influent is close to zero, f_H is the only connector parameter responsible for the variations of oxygen in the connector output.

Some influence on elemental balances is also exerted by f_{N1} and f_{N2} (fractions of active autotrophs), but negligible compared to f_H since the concentration of autotrophs in sewage is usually very small.

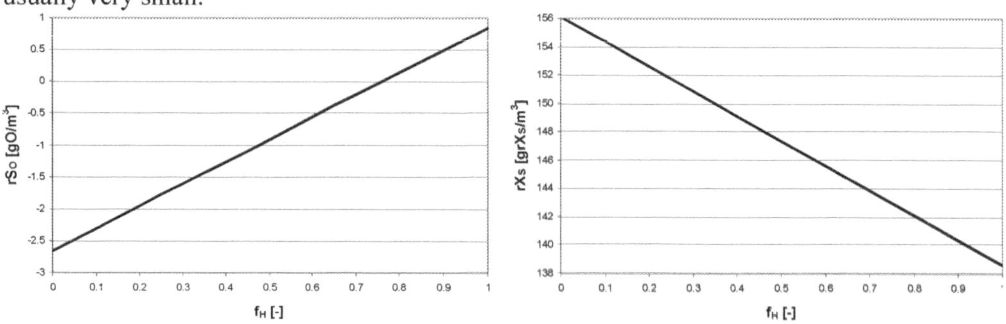

Figure 3.5 Influence of f_H on the most sensitive connector outputs, rS_O (left) and rX_S (right).

4 The river Lambro case study

An application of the connector has been made for the simulation of a 26 km long reach of the river Lambro, north of Milano, Italy. The most important pollution source in this stretch is the treatment plant of Merone. This treatment plant receives the wastewater of approximately 120.000 PE. Prior to an upgrade (+100% primary sedimentation and +50% biological treatment and secondary sedimentation volumes) the Merone WWTP was constantly working at its maximum hydraulic capacity (25.400 m^3/d), but this did not avoid the daily bypass of dry weather wastewater (approximately 40% was regularly discharged untreated into the river). The daily bypass of raw wastewater imposed a marked diurnal cycle in pollutant concentrations in the river for a considerable distance downstream of the plant.

The integrated system has been modelled with ASM1, a simplified version of RWQM1 (see (Vanrolleghem *et al.* 2001) for sub-model selection) and an *ad hoc* connector, all implemented in WEST®. Details on this case study can be found in (Benedetti *et al.* 2004).

One of the aims of the study was to estimate the impact of a plant upgrade on the receiving water quality. Simulation results concerning concentrations of ammonia and nitrates in the river in dry weather conditions can be seen in Figures 4.1 and 4.2 respectively, showing the plant impact before and after the upgrade. As expected from a nitrifying but not denitrifying plant, the nitrates concentration in the river increases and the ammonia concentration decreases.

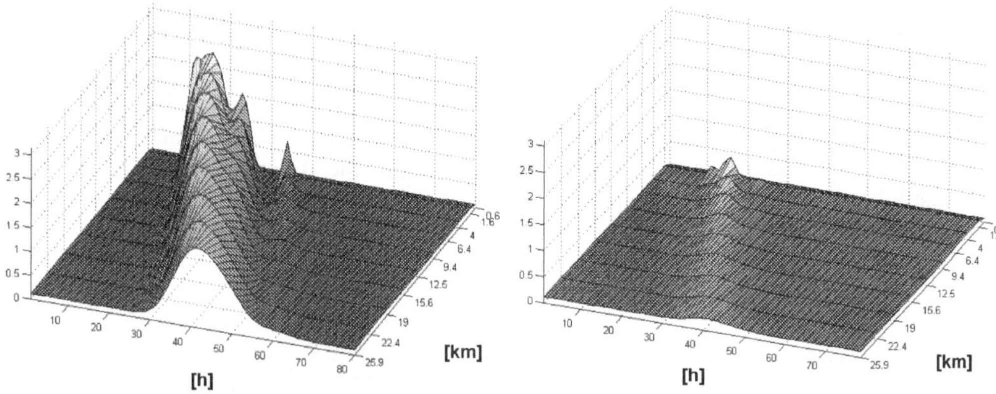

Figure 4.1 Ammonia concentration in the river in dry weather conditions before (left) and after (right) WWTP upgrade.

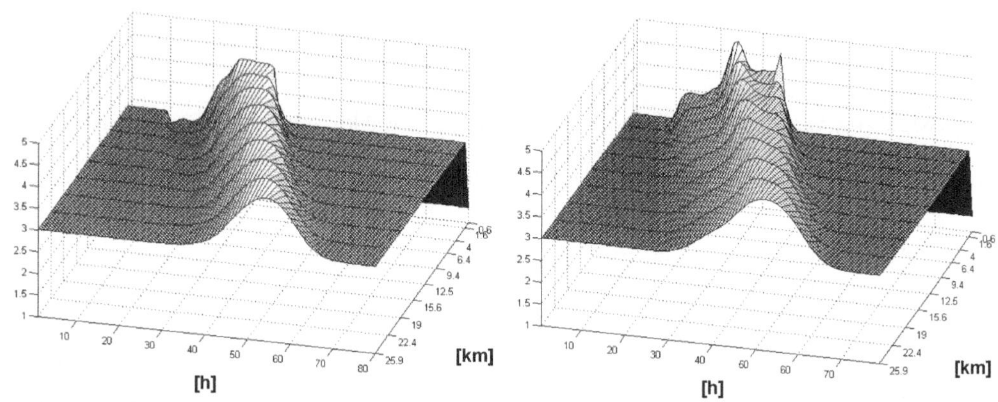

Figure 4.2 Nitrates concentration in the river in dry weather conditions before (left) and after (right) WWTP upgrade.

5 Conclusions

A connector between ASM1 and RWQM1 has been described in this paper. It should be considered as an example to introduce the approach to translate state variables from one model to another for integrated wastewater system modelling. Inherent features of this connector are its closed mass and elemental balances. The COD fractions of ASM1 have been split over the COD fractions of RWQM1, while compensation terms were used to close elemental balances. Also the different environmental conditions in the systems (activated sludge, sewer and river) have been taken into account to correct for inactivation of organisms when changing from one system to another. An evaluation of the influence of several connector and models parameters on connector compensation terms and outputs has been performed, showing that great importance lies in the elemental composition of the state variables, for which only scarce data can be found in literature. Some results of a case study have been described to show an application of such model connection approach, which is general and can be applied to any coupling of ASM-like models.

Acknowledgements

The results presented in this publication have been elaborated in the frame of the EU projects CD4WC, contract no EVK1-CT-2002-00118 and Harmoni-CA, contract no EVK1-CT-2002-20003. The programme is organised within the Energy, Environment and Sustainable Development Programme in the 5th Framework Programme for Science Research and Technological Development of the European Commission.

References

Benedetti, L., Meirlaen, J., Sforzi, F., Facchi, A., Gandolfi, C. and Vanrolleghem, P. A. (2004). Dynamic integrated modelling: A case study on the river Lambro. In: Proceedings of NOVATECH 2004, Lyon, France, 6-10 June 2004.

Brown, L. C. and Barnwell, T. O. (1987). The enhanced stream water quality model QUAL2E and QUAL2E-UNCAS: Documentation and user manual, U.S.EPA

Focht, D. D. and Verstraete, W. (1977). Biochemical ecology of nitrification and denitrification. In: Advances in microbial ecology. M. Alexander (ed.), Plenum Press, New York, pp. 135-214.

Fronteau, C., Bauwens, W. and Vanrolleghem, P. A. (1997). Integrated modelling comparison of state variables, processes and parameters in sewer and wastewater treatment models. Water Science and Technology 36 (5), 373-380.

Henze, M., Gujer, W., Mino, T. and van Loosdrecht, M. C. M. (2000). Activated sludge models ASM1, ASM2, ASM2d and ASM3. IWA, London.

Huisman, J. L., Krebs, P. and Gujer, W. (2003). Integral and unified model for the sewer and wastewater treatment plant focusing on transformations. Water Science and Technology 47 (12), 65-71.

Maryns, F. and Bauwens, W. (1997). The application of the Activated Sludge Model No.1 to a river environment. Water Science and Technology 36 (5), 201-208.

Meirlaen, J., Huyghebaert, B., Sforzi, F., Benedetti, L. and Vanrolleghem, P. A. (2001). Fast, simultaneous simulation of the integrated urban wastewater system using mechanistic surrogate models. Water Science and Technology 43 (7), 301-309.

Mourato, S., Matos, J., Almeida, M. and Hvitved-Jacobsen, T. (2003). Modelling in-sewer pollutant degradation processes in the Costa do Estoril sewer system. Water Science and Technology 47 (4), 93-100.

Reichert, P., Borchardt, D., Henze, M., Rauch, W., Reichert, P., Somlyódy, L. and Vanrolleghem, P. A. (2001). River Water Quality Model No. 1: II. Biochemical process equations. Water Science and Technology 43 (5), 11-30.

Shanahan, P., Borchardt, D., Henze, M., Rauch, W., Reichert, P., Somlyódy, L. and Vanrolleghem, P. A. (2001). River Water Quality Model No. 1: I. Modelling approach. Water Science and Technology 43 (5), 1-9.

Spanjers, H., Vanrolleghem, P. A., Nguyen, K., Vanhooren, H. and Patry, G. G. (1998). Towards a simulation-benchmark for evaluating respirometry-based control strategies. Water Science and Technology 37 (12), 219-226.

Vanhooren, H., Meirlaen, J., Amerlink, Y., Claeys, F., Vangheluwe, H. and Vanrolleghem, P. A. (2003). WEST: Modelling biological wastewater treatment. J. Hydroinformatics 5 (1), 27-50.

Vanrolleghem, P. A., Borchardt, D., Henze, M., Rauch, W., Reichert, P., Shanahan, P. and Somlyódy, L. (2001). River Water Quality Model No. 1: III. Biochemical submodel selection. Water Science and Technology 43 (5), 31-40.

Combined Sewer Overflows Into The Crati River (Cosenza, Italy)

G. Palma*, F. Calomino* and P. Piro*

* Dipartimento di Difesa del Suolo "V. Marone", Cubo 42B - Università della Calabria, 87036 Arcavacata di Rende, ITALY.
(e-mail: gpalma@dds.unical.it)

Abstract Combined Sewer Overflows (CSOs) have been for long time present in the technical literature. A short review of recent contributions is given in the paper. Regulations for reducing CSO impact in the U.S., EU and in Italy are then considered. Different means of controlling CSOs are discussed and among them retention storage is considered as the most frequent. Observations from an experimental catchment in Cosenza, Italy, drained by a combined sewer system with overflow facility, are then presented. They include 28 CSO events, for which the complete hydrographs were registered, obtaining the overflows hydrographs. Six more events were observed for quality data, and TSS, BOD and COD were measured; a regression equation between TSS, BOD and COD was found. Pollutant rates were represented against stormwater volumes, and first foul flush seems to be present only in one of the six events. As EU regulations demand complete storage and treatment of combined sewer overflows, different retention volumes of an off-line tank were analysed, provided that a discharge from 3 to 4 times the dry weather discharge is continuously treated by the treatment plant during storms. Values of the order of 150-200 m^3/ha should reduce the final spill into receiving waters to a few events per year.

Keywords Combined sewer overflows, urban catchment, wastewater, retention capacities

1. Combined Sewer Overflows – A Rapid Overview

Combined sewers are diffused all over the world and especially in cities where sewerage systems were developed during centuries, like in Europe.

Musso and Chocat (1998) point out that combined sewers are the most common solution and that Combined Sewer Overflows (CSOs) are a danger for water bodies. As reported by Ashley and others (2003), in Denmark there are some 7000 CSO structures, many of which will be abandoned or supplemented by detention basins. Blanksby (2002) reports that about 7000 unsatisfactory CSOs should be improved until 2010 in U.K. According to Fujita (1998), the 85% of metropolitan area of Tokyo is mainly served by combined sewers.

There are many experimental observations in catchments served by combined sewers. The ASCE – WEF Manual of engineering practice (1992) reports the following observed pollutant concentrations in CSOs:

 Sewer Networks and Processes within Urban Water Systems edited by J.-L. Bertrand-Krajewski, M. Almeida, J. Matos & S. Abdul-Talib. ISBN: 1 84339 506 1.

Table 1. CSOs observed pollutant concentrations (ASCE-WEF, 1992)

	TSS (mg/l)	BOD (mg/l)	COD (mg/l)
Range	273-551	59-222	264-481
Average	370	115	367

In Italy, recent observations carried out in the city of Pavia (Papiri, 2003) and concerning an experimental catchment of area 11 ha, served by a combined sewer, point out the following values:

Table 2. Combined sewer observed pollutant concentrations in the city of Pavia (Papiri, 2003)

	TSS (mg/l)	BOD (mg/l)	COD (mg/l)
Range	20-2360	8-1780	28-2434
Average	502	281	484

The above values show that combined sewer pollutant concentrations can be by far above those allowed for discharge into superficial waters; indeed there are many countries in which CSOs are the object of specific regulations, aiming at reducing the pollution into receiving water bodies.

2. Regulations In USA and EU

Combined sewer overflows were traditionally allowed, provided that the dilution rate (discharge of wastewater divided by discharge of stormwater) was of the order of 1/2 to 1/5. The problem of reducing CSOs historically arose in situations where use of water bodies had a social or economic value: protection of drinking water resources, protection of bathing, etc.

During the mid 1970s and early 1980, as reported by Roesner and Rowney (1996), it became increasingly apparent that pollution in wet weather discharges was significant and national regulations began to be issued. In the U.S., in particular, the National CSO Control Policy was finalized in 1994, based on three principles: 1) an accurate characterization of the CSO system, 2) demonstration of implementation of nine minimum controls, that can be implemented in a relatively short period (e.g., less than approximately two years), and 3) development and implementation of a long-term CSO control plan. The nine minimum controls are also reported in detail by the Authors.

The European countries belonging to the EU follow with their national legislation the EU Council Directive of 21 May 1991 concerning urban waste water treatment (91/271/EEC). According to its definitions, "domestic waste water" means waste water from residential settlements and services which originates predominantly from the human metabolism and from household activities; "industrial waste water" means any waste water which is discharged from premises used for carrying on any trade or industry, other than domestic waste water and runoff rain water; and "urban waste water" means domestic waste water or the mixture of domestic waste water with industrial waste water and/or runoff rain water. Secondary treatment is the standard for urban waste water disposal into superficial waters.

Following the Directive, European countries have adopted national regulations for CSO control.

Among others, France adopted the Decree 22 Dec. 1994, Portugal adopted the Decree 74/090, Germany adopted the ATV-a 128 and ATV-A 138, and Italy adopted the Decrees 152/99 and 258/00. In UK, as reported by Blanksby (2002), the consent, based on performance measures suggested by the Environment Agency's Consent Manual in 1997, is a legal document which sets out specific conditions under which the CSO must operate.

3. CSO CONTROL

CSO control is recognized as a necessity; nevertheless, the means of operating such a control are technically different. The ways of reducing loads discharged by of CSOs in the USA consist traditionally in on-line or off-line retention storage (ASCE-WEF 1992). Also in Japan the main solution consists in stormwater storage and treatment in wastewater plants (Fujita, 1998). In Germany, CSDT (Combined Sewer Detention Tanks) are the most common way of dealing with combined sewers in wet weather and storage facilities are considered an essential part of the sewer system (Kowalski et al., 1998). In France, many catchments are served by stormwater reservoirs, as reported by Faure et. al (1998) for the city of Nancy, by Perez-Sauvagnat et al. (1998) for the Seine st. Denis, and by Charry and Lussagnet (1998) for the city of Marseille.

According to several authors (Ashley et al., 2003), the sediments carry most of the pollution loads (30% to 50%). Many aspects of the mechanics of suspended transport and that of near-bed transport are still under study. In particular the weak resistance to shear causes rapid re-entrainment of deposited solids with the increase of discharge. This fact, especially for combined sewers, influences greatly the impact of CSO on receiving waters.

Retention capacities are the most used device to reduce pollutant loads; nevertheless their behaviour is known mainly under an empirical approach. Recently, mechanics of sedimentation processes in ponds is taken into consideration. Marsalek et al. (1998, 2003) show the distribution of suspended solids in a stormwater pond. Other facilities, like vortex separators, can be used, with the aim of reducing CSOs pollutant loads, in association or not to stormwater storage.

Balmforth et al (1998) show the use of grids and special spillways to comply with the requirements of EU directive about surface water pollution by CSOs. There are three standard types of spillways used in U.K.: high side weir, extended stilling pond and vortex with peripheral spill.

Luyckx et al. (1998) show the dimensioning of a high side weir to optimise the separation of a fraction of sediments. Tyack and Fenner (1998) show the hydraulics of vortex separators, that are increasingly used in the sewer network and at the wastewater treatment plants. Such devices are considered ideal for heavy suspended solids removal.

The impact on treatment plants is a complex problem, taken into consideration among others by Dormoy et al. (1998); they conclude that usual plants need special opportunities like the use of stormwater tanks, allowing the increasing of sludge production, O_2 requirement and also sludge quality.

Detention storage is the practice that allows to detain part of the runoff volume in capacities inside the catchment and it is frequently associated with local runoff control. These possibilities are interesting, but their use under complete control is questionable (Schilling, 2003).

Real time control (Campisano and Modica, 2002) allows the best use of in-sewer capacities, by using mobile devices and controls. Even though this technique is promising, for the time being the examples known in practice only concern a few cases.

It is very likely that, for years on, off-line and on-line retention storage will remain the practical way of controlling CSOs.

4. THE EXPERIMENTAL CATCHMENT IN COSENZA (ITALY)

During many years University of Calabria studied the urban catchment of the Liguori creek in the city of Cosenza, Italy. The area of the catchment is 414 ha in total, consisting of buildings (10.2%), gardens (0.8%), pavements and roads (37.6%) and a wide natural area (51.4%). Originally, the main drain was a natural stream; today, because of the increasing urbanisation, its urban reach has become a combined sewer main with a concrete horseshoe section. Upstream to its outfall into the Crati river, a sewer collector intercepts the wastewater and conveys it to the treatment plant.

Three experimental stations for data collection are in use at the catchment; the one considered for the aim of the present work is located at the catchment outlet (station 1) and includes a rain gauge, an ultrasonic level gauge and an automatic sampler. Further information concerning measurement stations and catchment can be found in the literature (Calomino et al., 2000; Piro and Sole , 2001).

Following figure 1 shows the catchment and the sewer network.

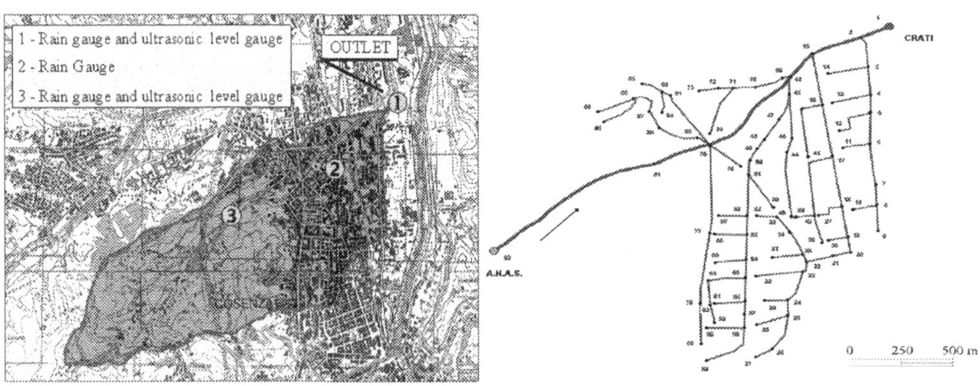

(a) (b)

Figure 1. Experimental catchment (a) and sewer network (b)

The overflow structure was studied by means of a physical model in Froude similarity (Calomino et al., 2003). When the spillway is operating, the flow derived to the treatment plant is almost constant and equal to 650 l/s, that is about 3-4 times the dry-weather discharge. Over 120 rain-runoff events were observed. For each of these events the hydrographs passing through the overflow structure and discharged into the Crati river are available.

For the purpose of sizing a retention storage, 28 significant events observed from december 1994 to may 1996 were selected, and the relative overflows were computed. The following Table 3 shows the rain-runoff characteristics and the overflows of these events.

Table 3. 28 Events observed between Dec. 1994 and May 1996; h is the total rainfall depth, V_T and V_{OF} are, respectively, the total volume and the overflow volume.

Event n.	Date	Duration (min)	h (mm)	V_T (m^3)	V_{OF} (m^3)	V_{OF}/V_T (%)
1	21/12/94 19.50	250	13,20	10056	2677	26,6
2	22/12/94 19.05	295	56,80	53584	30862	57,6
3	24/12/94 17.30	391	41,20	30588	9701	31,7
4	1/1/95 22.30	90	52,40	57291	35798	62,5
5	3/1/95 7.00	511	34,00	33557	14346	42,8
6	7/1/95 0.50	281	8,40	8757	2073	23,7
7	12/1/95 2.36	175	8,32	6055	1277	21,1
8	22/1/95 0.16	125	6,40	4501	844	18,8
9	31/1/95 3.00	241	28,00	30676	9022	29,4
10	14/2/95 17.30	361	24,40	12511	3157	25,2
11	16/2/95 19.00	211	15,20	10197	4762	46,7
12	18/2/95 19.00	211	28,00	24117	5036	20,9
13	25/2/95 4.15	541	42,80	43009	21587	50,2
14	2/3/95 8.20	641	96,40	98968	47478	48,0
15	9/3/95 6.15	286	37,60	38023	18631	49,0
16	28/3/95 10.15	191	31,52	24144	12976	53,7
17	29/3/95 12.50	551	44,80	54672	13323	24,4
18	9/4/95 3.05	176	10,80	9215	4596	49,9
19	26/4/95 1.00	211	63,20	42582	24712	58,0
20	2/5/95 11.21	420	24,32	18681	6117	32,7
21	14/5/95 6.00	211	11,20	9276	2491	26,9
22	1/8/95 2.10	261	14,80	12132	5483	45,2
23	16/8/95 9.55	96	21,60	11300	7869	69,6
24	24/8/95 18.30	91	10,00	2912	1142	39,2
25	25/8/95 21.20	161	14,40	11291	5537	49,0
26	6/2/96 5.23	218	24,80	59623	28691	48,1
27	30/4/96 20.00	240	48,00	41435	16773	40,5
28	11/5/96 10.15	361	27,60	25100	11721	46,7

Dry weather discharges are ranging between 150 and 250 l/s. During dry weather the analysis showed the following values of BOD5, COD, TSS, reported in tab. 4 (Calomino et al., 2000).

Table 4. Analysis during dry weather.

	BOD_5 (mg/l)	COD (mg/l)	TSS (mg/l)
Range	12.5 – 185	21 – 392	1.3 – 279
Mean Value	87	170	87

5. STORMWATER QUALITY

For the quality of stormwater concern, the paper takes into consideration six events. For each of them the following parameters were observed: BOD_5, COD, total suspended solids (TSS), pH and conductivity. Table 5 shows the runoff volume, range and average values recorded for each pollutant.

Table 5. Analysis during wet weather

		Range (mg/l)			average values (mg/l)		
Event	V_T (m^3)	TSS	COD	BOD_5	TSS	COD	BOD_5
24 MAR 98	11.222	27-344	144-157	15-250	165	150	66
13 APR 98	1.919	224-296	273-354	150-160	251	304	155
28 APR 98	11.568	73-342	42-760	-	197	453	-
1 NOV 00	6.472	141-1333	160-1387	152-544	574	583	578
24 FEB 02	10.599	32-620	46-684	4-201	195	229	80
23 MAR 02	2.673	116-193	212-310	95-125	159	267	116

In total, 84 samples were analysed. A regression analysis was carried out for BOD_5, COD and TSS, with the results reported in the following table 6.

Table 6. Regression analysis among quality parameters

Regression Equations	R^2
COD = 0,76 TSS+143,77	0,79
BOD_5 = 0,36 TSS+27,87	0,83

Next figure 2 shows the correlation between COD and TSS.

The pollutographs showed that pollutant concentrations have the same general pattern as sewer flow, Q, and that concentration peaks correspond generally to discharge peaks of each rainfall event (Calomino et al., 2003).

Following figure 3 shows the ratio between measured mass and total mass (M/M_T) versus the ratio between volume and total event volume (V/V_T). Only in the event of 24 MAR 98 a first foul flush appears clearly.

On the other hand, average concentrations of pollutants are well above those typical of dry-time and those admitted for receiving superficial waters.

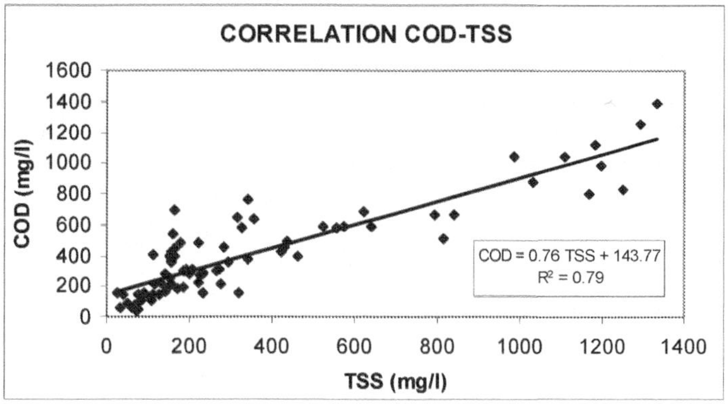

Figure 2. - Regression COD – TSS

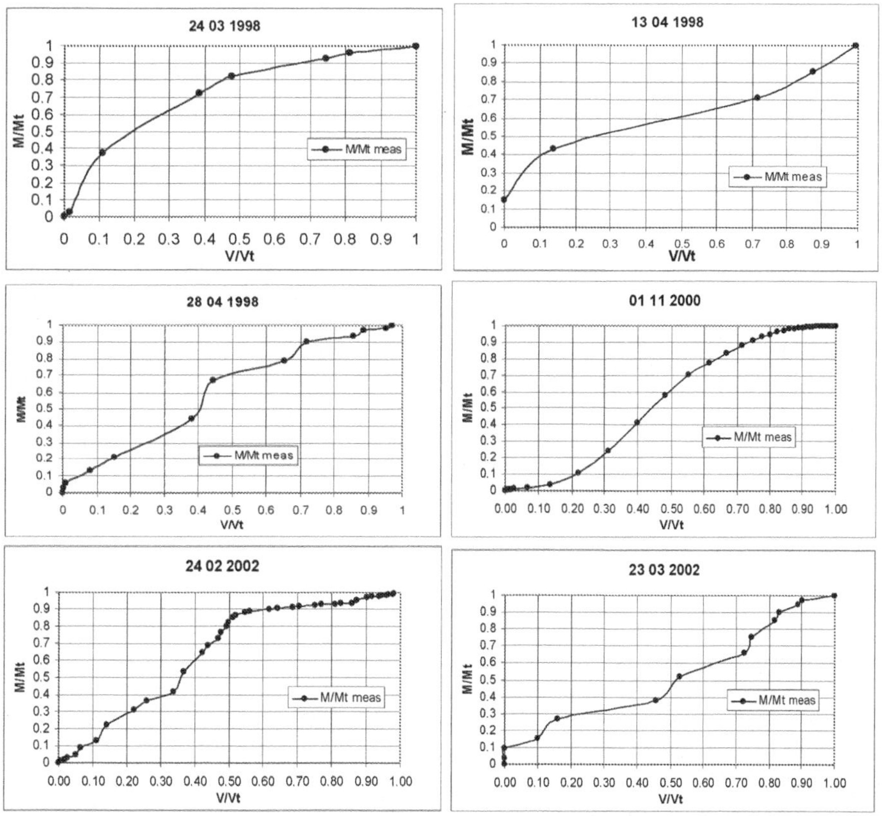

Figure 3. M/M_T versus V/V_T values

6. RETENTION STORAGE CAPACITY

According to EU Directive, uncontrolled discharge of combined stormwater in receiving waters must be avoided, so it's necessary providing a storage capacity and an adequate treatment. It is clear that this demands considerable retention capacities. According to Paoletti et. al (1998) for the italian climate capacities up to 200 m^3/ha are needed.

For the case under examination, we must consider that anyway a considerable amount of the runoff volume V_T is derived to the treatment plant. For the 28 events in tab. 1, one can see that the ratio V_{OF}/V_T is ranging between 30% and 70%, with an average of about 40%. This allows a considerable reduction of overflow volumes.

If we assume an off-line retention tank of capacity V_R, the overflow volume V_{OF} is collected into it; once this capacity is exceeded, the volume $V_D = V_{OF} - V_R$ is spilled into the receiving body.

It is known that there are different ways of assessing the performance of retention facilities (Bertrand-Krajewski et. al, 2002; Lau et al, 2002).

In this paper were considered only the following indicators, for each retention volume:

- Maximum values of discharged volume $V_{D,MAX}$, and corresponding rain depth, $r_{d,MAX}$
- Maximum ratio between discharged volume and overflow volume $(V_D/V_{OF})_{MAX}$
- Sum of discharged volume divided by sum of overflow volume $(\Sigma V_D/\Sigma V_{OF})$
- Number of overflow events, n_e

The obtained values are reported in the following table 7.

Table 7. CSOs indicators for Liguori creek

V_R (m^3/ha)	50	100	150	200	250
$V_{D, MAX}$ (m^3)	37.478	27.478	17.478	7.478	0
$r_{d, MAX}$ (mm)	19	14	9	4	0
n_e	12	6	3	1	0
$(V_D/V_{OF})_{MAX}$	0.8	0.6	0.4	0.2	0
$\Sigma V_D/\Sigma V_{OF}$	0.45	0.16	0.12	0.07	0

7. CONCLUSIONS

It is difficult and very expensive to avoid completely CSOs into receiving waters. Indeed, EU directive 271/91 suggests the adoption of measures to limit pollution from storm water overflows, based on dilution rates or capacity in relation to dry weather flow, or a certain acceptable number of overflows per year.

From the observed hydrographs we can conclude that retention capacities of 50-100 m^3/ha look inadequate, while capacities of 150-200 m^3/ha should be chosen to reduce the number of spills to receiving waters to 1-3 per year, provided that a discharge of 3-4 times the dry-weather flow is continuously derived during storms to the treatment plant, and overflow volumes can be consequently reduced. In this case, maximum rain depths of the order of 9 to 4 mm, (40% to 20% of the runoff volume) are disposed in the receiving river during a storm. On a long-period basis, about 10% of the runoff volume of the observation period is disposed into receiving waters without any treatment.

REFERENCES

ASCE – WEF (1992). Design and Construction of Urban Stormwater Management Systems. ASCE Manuals and Reports of Engineering Practice No. 77. WEF Manual of Practice FD-20. ISBN 0-87262-855-8.

Ashley, R.M., Sakrabani, R., Hvitved-Jacobsen, T. and Vollerstsen, J. (2003). Solids in Sewers. Near bed transport and the biodegradability of eroded solids. In *Proc. of the XXX IAHR Congress,* AUTh, Thessaloniki, Greece, August 2003, Theme B, 409-416. ISBN 960-243-594-1.

Balmforth, D., Blanksby, J., Brown, C. and Rogerson, C. (1998). Development and testing of a new design procedure for the control of aesthetic pollutants at combined sewer overflows. In *Proc. of Novatech, 3rd Int. Conference*, vol. 1, 339-346. ISBN 2-9509337-1-8.

Bertrand-Krajewski, J.-L., Barraud, S. and Bardin, J.-P. (2002). Uncertainties, performance indicators and decision aid applied to stormwater facilities. *Urban Water* **4**, 163-179.

Blanksby, J. (2002): A review of the UK approach to measuring the performance of combined sewer overflows. *Urban Water* **4** , 191-198.

Calomino, F., De Rosa, S., Frega, G., Piro, P., Veltri, P., Mauro, M. and Pranno, A. (2000). Water quality observations in an urban experimental basin. In *Proc. of New trends in water and Environmental engineering for safety and life. International Conference*, Capri (Italy), July 03-07. Balkema, Rotterdam. ISBN 90-5809-138-4.

Calomino, F., Piro, P., Frega, F and Palma, G. (2003). Sversamento di acque miste in un corso d'acqua ricettore. Il caso del canale Liguori a Cosenza. In *Atti del 24° Corso di Aggiornamento in Tecniche per la Difesa dall'Inquinamento*, 623-644. Editoriale BIOS, Cosenza. ISBN 88 7740 373 X.

Campisano, A.P. and Modica, C. (2002). Modern Technologies for real time control of wastewater systems. In *Proc. of the 16th European Junior Scientist Workshop "Real time control and measurements in urban drainage systems"*. Catania, 7-10 Nov., 3-15. Editor Campisano, A.

Charry, J.P. and Lussagnet, P. (1998). Rain water treatment "Grand littoral" Site Marseille (France). In *Proc. of Novatech, 3rd Int. Conference*, vol. 2, 53-58. ISBN 2-9509337-2-6.

Dormoy, T., Tisserand, B. and Herremans, L. (1998). Incidence des apports de temps de pluie sur les contraintes d'exploitation d'une usine de dépollution. In *Proc. of Novatech, 3rd Int. Conference*, vol. 2, 83-90. ISBN 2-9509337-2-6.

Faure, D., Schmitte, J.P. and Badot, R. (1998). Réduction des déversements de pollution par temps e pluie par une stratégie globale pollution/inondation de gestion d'un réseau d'assainissement. In *Proc. of Novatech, 3rd Int. Conference*, vol. 2, 287-294. ISBN 2-9509337-2-6.

Fujita, S. (1998). Toward "Zero Emissions" in urban drainage works. In *Proc. of Novatech, 3rd Int. Conference*, vol. 1, 331-338. ISBN 2-9509337-1-8.

Kowalski, R., Reuber, J. and Kongeter, J. (1998). Investigation and Optimization of the Performance of Sewage Detention Tanks during CSO. In *Proc. of Novatech, 3rd Int. Conference*, vol. 1, 83-90. ISBN 2-9509337-1-8.

Lau, J., Butler, D. and Schutze, M. (2002). Is combined sewer overflow spill frequency/volume a good indicator of receiving water quality impact?. Urban Water **4**, 181-189.

Luyckx, G., Vaes, G. and Berlamont, J. (1998). Experimental investigation on the efficiency of a high side weir overflow. In *Proc. of Novatech, 3rd Int. Conference*, vol. 1, 347-354. ISBN 2-9509337-1-8.

Marsalek, J., Krishnappan, B.G., Watt, W.E. and Anderson, B.C. (1998). Size distribution of suspended sediments in an on-stream stormwater management pond. In *Proc. of Novatech, 3rd Int. Conference*, vol. 1, 543-550. ISBN 2-9509337-1-8.

Marsalek, J. and Krishnappan, B.G. (2003). Sediment settling in stormwater management ponds: current research and practice. In *Proc. of the XXX IAHR Congress,* AUTh, Thessaloniki, Greece, August 2003, Theme B, 417-424. ISBN 960-243-594-1.

Musso, E. and Chocat, B.(1998). Evaluation des rejets urbains de temps de pluie de grands bassins versants. Contribution méthodologique à la comparaison des outils et à l'évaluation des incertitudes. Application à la communauté urbaine de Lyon. In *Proc. of Novatech, 3rd Int. Conference*, vol. 1, 75-82. ISBN 2-9509337-1-8.

Paoletti, A., Mambretti, S., Sanfilippo, U. and Brown, M. (1998). The EU-SPRINT 098 Baggio experimental pilot study for CSO volume and frequency reduction. In *Proc. of Novatech, 3rd Int. Conference*, vol. 2, 263-270. ISBN 2-9509337-2-6.

Papiri, S. (2003). Qualità delle acque meteoriche di dilavamento, con particolare riferimento alle superfici stradali, e tecniche di controllo. In *Atti del 24° Corso di Aggiornamento in Tecniche per la Difesa dall'Inquinamento*, 579-600. Editoriale BIOS, Cosenza. ISBN 88 7740 373 X.

Perez-Sauvagnat, I., Maytraud, T. and Piel, C. (1998). L'ouverture au public de bassins de retenue en site urbain: un pari difficile mais nécessaire. In *Proc. of Novatech, 3rd Int. Conference*, vol. 1, 313-320. ISBN 2-9509337-1-8.

Piro, P. and Sole, A. (2001). Analisi di reti di drenaggio urbano mediante l'uso dei GIS. Un'applicazione al bacino urbano del Liguori (CS). In *L'uso dei GIS per l'analisi e il controllo di problematiche territoriali*, 53-84. Editoriale BIOS, Cosenza. ISBN 88-7740-320-9

Roesner, L.A and Rowney, C. (1996). National storm water quality regulations and standards. *Journal of Hydraulic Research* **34**, 841-856.

Schilling, W. (2003). Urban drainage – Quo Vadis?. In *Proc. of the XXX IAHR Congress,* AUTh, Thessaloniki, Greece, August 2003, Theme B, 417-424. ISBN 960-243-594-1.

Tyack, N.J. and Fenner, R.A. (1998). Identification of flow regimes within a hydrodynamic separator. In *Proc. of Novatech, 3rd Int. Conference*, vol. 1, 361-368. ISBN 2-9509337-1-8.

Comparison between the performances of two combined sewer overflow devices in the reduction of water volume and pollutant discharges

A.P. Campisano*, E. Creaco and C. Modica*****

* Dipartimento di Ingegneria Civile e Ambientale, Università di Catania, Viale A. Doria 6, 95125 Catania, Italy (E-mail: acampisa@dica.unict.it)
** Dipartimento di Ingegneria Civile e Ambientale, Università di Catania, Viale A. Doria 6, 95125 Catania, Italy (E-mail: ecreaco@dica.unict.it)
*** Dipartimento di Ingegneria Civile e Ambientale, Università di Catania, Viale A. Doria 6, 95125 Catania, Italy (E-mail: cmodica@dica.unict.it)

Abstract Storm water is a significant source of pollution in urban catchments, being the vehicle of both conventional pollutants and toxic substances. The growing attention towards environmental problems has consequently determined the necessity of adopting proper combined sewer overflow (CSO) devices able to reduce the pollutant load into receiving waters during rain events.
In this paper, one of these devices – the high crested side-weirs with bottom orifice – is analysed. In particular a comparison, in terms of reduction of discharged water volumes and pollutants, between the performances of the high crested CSO device and the more common low crested side-weir is presented.
For this purpose, numerical simulations have been run using experimental data concerning flow rates and pollutant concentrations measured during different rain events in the urban catchment of Cascina Scala, Italy.
A model based on the solution of the fully-dynamic De Saint Venant equations for the water flow and of the advection-dispersion equation for the propagation of pollutants inside the sewer has been implemented and applied. The simulations show the better performances of the analysed high crested CSO device compared to the low crested side-weir.

Keywords Combined sewer overflows; side-weirs; pollutant control.

Introduction

As well known, the objective of combined sewer overflow (CSO) devices is to discharge exceeding flow rates into water bodies when a sufficient pollutant dilution is achieved during rain events. For this purpose, many kinds of CSO devices, such as the side-weir, the leaping and the baffled weir, have often been used with the main function of flow partition.

In the last decade, many authors (Xanthopoulos and Hahn, 1993; Deletic and Maksimovic, 1998; Gray and Becker, 2002) have shown that storm water can be a significant source of pollution in urban catchments, being the vehicle of both conventional pollutants (e.g. suspended solids, COD, BOD) and heavy metals (e.g. lead, cupper, zinc). Additionally, long dry periods followed by one or more storm events can determine relevant problems of first flush pollution, due to the re-suspension of sediments deposited on the catchment surfaces and on the bottom of the sewer collectors.

 Sewer Networks and Processes within Urban Water Systems edited by J.-L. Bertrand-Krajewski, M. Almeida, J. Matos & S. Abdul-Talib. ISBN: 1 84339 506 1.

The great concern for the environmental problems has consequently induced to consider the CSO discharges also from the qualitative point of view and to carry out the design and set-up of proper CSO devices. In particular, more efficient devices, compared to the traditional low crested side-weir (Figure 1a), have been studied by many authors in the last decades. Some of these, such as the high crested side-weir with bottom orifice (Artina *et al.*, 1997), the sewer overflow chamber (Saul and Delo, 1982) and many kinds of swirl and vortex devices (Smisson, 1967; Field, 1972; Brombach, 1992; Andoh and Saul, 2002) allow both quantity and quality control of stormwater discharges. In fact, these devices can perform a relevant reduction of the discharged pollutant loads, besides the usual flow partition function (Field, 1990).

In particular, the high crested side-weir with bottom orifice (Figure 1b) is made up of a side weir located in a higher position than in the commonly used side-weirs and of a downstream fixed gate with bottom orifice.

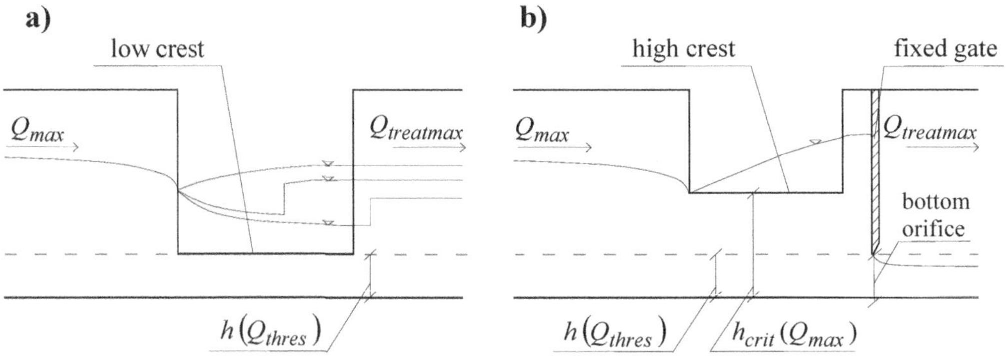

Figure 1 a) The low crested side-weir; b) The high crested side-weir with bottom orifice.

As well known, the crest level of the usual low crested CSO device is calculated in steady flow conditions with reference to a threshold flow rate value Q_{thres}, corresponding to the beginning of the discharge process, while the side weir length is related to the maximum incoming flow rate Q_{max} and to the maximum treatment plant flow capacity $Q_{treatmax}$.

Instead, the crest level of the high crested CSO device is sized in steady flow conditions as function of the critical depth h_{crit} relative to Q_{max}, being the side weir length still calculated as function of the same Q_{max} and of $Q_{treatmax}$. For this device, the regulation of the flows addressed to the treatment plant is assigned to the bottom orifice, which is designed with reference to the value of Q_{thres}. In particular, when flow rates exceed the threshold value, the bottom orifice allows for activating the collector in-line storage capacity upstream the gate. The spilling process begins only when the water level upstream the gate rises up over the weir crest, reducing the discharge of first flush waters and of pollutants into the receiving water body. In addition, in comparison to the low crested side-weirs, the high crested CSO device should increase the water volumes conveyed to the treatment plant.

For both the devices, during the normal functioning, the flow rate conveyed to the treatment plant, equal to the difference between the incoming flow rate and the discharged flow rate, always results greater than the threshold value, according to the device efficiency.

In this paper, a comparison between the behaviour of the two CSO devices described in figure 1 is presented. In particular, performances are compared in terms of reduction of the discharged volumes and pollutants.

For this objective, a numerical model was set up and applied to the experimental data of the urban catchment of Cascina Scala, Italy (Ciaponi *et al.*, 2002), using both quantitative and qualitative measurements recorded during different rain events in the period 2000-2001.

The implemented model is based on the solution of the fully-dynamic De Saint Venant equations for the water flow and of the advection-dispersion equation for the analysis of the pollutants inside the sewer system.

Methods

The adopted model

For a prismatic collector, the 1-D De Saint Venant equations for the unsteady flow and the advection-dispersion equation for the pollutant propagation can be written in the following conservative-law form:

$$\frac{\partial \mathbf{U}}{\partial t}+\frac{\partial \mathbf{F}(\mathbf{U})}{\partial x}=\mathbf{D}(\mathbf{U}), \tag{1}$$

where the terms $\mathbf{U}$ (dependent variable vector), $\mathbf{F}(\mathbf{U})$ (flux vector) and $\mathbf{D}(\mathbf{U})$ (source-diffusive term vector) can be written as follows:

$$\mathbf{U}=\begin{pmatrix} A \\ Q \\ AC \end{pmatrix}, \qquad \mathbf{F}(\mathbf{U})=\begin{pmatrix} Q \\ VQ+F_h/\rho \\ QC \end{pmatrix}, \qquad \mathbf{D}(\mathbf{U})=\begin{pmatrix} -q_o \\ gA(i-J)-Vq_o \\ \dfrac{\partial}{\partial x}\left(AD\dfrac{\partial C}{\partial x}\right)-Cq_o \end{pmatrix} \tag{2}$$

where x [m] and t [s] are space and time independent variables; A [m^2] is the water cross-section; Q [m^3/s] and V [m/s] are the flow rate and the flow velocity respectively; F_h [N] is the hydrostatic force over the cross section; ρ [kg/m^3] is the water density; g [m/s^2] is the gravity acceleration; i is the bed slope; J is the friction slope; q_o [m^2/s] is the lateral outflow discharge per unit length of collector; C [kg/m^3] is the pollutant mass concentration and D [m^2/s] is the dispersion coefficient.

The friction slope has been evaluated using the Strickler relation:

$$J=\frac{V^2}{k^2R^{4/3}}, \tag{3}$$

being R [m] and k [m$^{1/3}$/s] the hydraulic radius and the collector roughness coefficient respectively.

The lateral outflow discharge from the side-weir has been evaluated using the following relation:

$$q_o=\mu\sqrt{2g}(h-b)^{3/2}, \tag{4}$$

where h [m] is the water level, b [m] is the height of the side-weir crest and μ is the discharge coefficient along the side-weir assumed equal to 0.38 as usual.

For the gate orifice outflow, the following relation has been used:

$$Q = C_c \, a \, B \sqrt{2g\left(h_{up} + \frac{V_{up}^2}{2g} - h_{dw}\right)}, \tag{5}$$

where a [m] is the gate opening; B [m] is the gate width; h_{up} [m] and h_{dw} [m] are the water levels upstream and downstream the gates; V_{up} [m/s] is the upstream flow velocity. The contraction coefficient C_c is derived from Ghetti (1984) as:

$$C_c = 0.61 + 0.39\left(\frac{a}{h_{up}}\right)^{4.836}, \tag{6}$$

The pollutant dispersion coefficient D has been calculated using the following relation, theoretically derived by Elder (1959) for turbulent shear flow conditions:

$$D = 5.93 h u^*, \tag{7}$$

where u^* [m/s] $= \sqrt{gRJ}$ is the friction velocity. For sewer systems, equation (7) generally provides values of D within the range 0.5-2.5, according to the values suggested by many authors (Payne *et al.*, 1990; Garsdal *et al.*, 1995; Mark *et al.*, 1998; Schlütter and Mark, 2002) for the propagation of pollutants in sewer collectors.

Considering low concentrations, water flow is assumed not to be influenced by the pollutant propagation phenomena. Therefore, equation (1) can be solved following a totally uncoupled approach (Armanini, 1999). In particular, the numerical solution of the De Saint Venant equations allows for determining the hydraulic variables used for the resolution of the advection-dispersion equation at the same time-step.

The second order predictor-corrector McCormack numerical scheme (Garcia-Navarro and Saviron, 1992) was adopted for solving equation (1); an additional step based on the theory of the Total Variation Diminishing (TVD) (Garcia-Navarro *et al.*, 1992) has been applied for reducing the numerical oscillations due to the water level and flow rate high gradients.

Boundary conditions are imposed to the hydraulic variables Q and A at the upstream and downstream collector ends. In particular, an entering hydrograph $Q(t)$ is considered at the upstream boundary and a relation $Q(h)$ is prescribed at the downstream end depending on the local flow conditions. Water cross-sections A are correspondingly evaluated by applying the method of characteristics (Garcia-Navarro and Saviron, 1992). The entering pollutograph $C(t)$ is also imposed at the upstream collector end, as boundary condition necessary for the solution of the advection-dispersion equation.

The model has been recently validated for what concerns both the hydraulic part and the pollutant transport part. In particular, the hydraulic part was tested against experimental data derived from laboratory and field experiments in rapidly varied flow conditions (Bertrand-Krajewski *et al.*, 2004; Campisano *et al.*, 2004); the pollutant transport part, instead, was tested against the analytical solutions of the advection-dispersion equation in simplified conditions (Ogata and Banks, 1961).

Description of the system

The urban catchment of Cascina Scala (Ciaponi *et al.*, 2002) is a residential district located on the northern outskirts of Pavia, Italy. The total contributing area adds up to 113480 m^2 (65% impervious, 35% pervious and not connected to the sewer system) and the number of inhabitants is about 3000. An average flow rate Q_d of 0.007 m^3/s is conveyed in the sewer system during dry period (Papiri, 2000). The final collector of the considered urban catchment is characterised by a 70x105 cm egg-shaped cross section and an average slope of about 0.45% (Ciaponi and Papiri, 1994). A collector with a simpler 60x100 cm rectangular cross section, with a wetted area equivalent to the real one (for filling degree of 80%) and a total length of 500 m has been considered for the numerical simulations. Assuming a Strickler coefficient $k = 70$ $m^{1/3}/s$, a maximum flow capacity Q_{max} of about 0.85 m^3/s has been obtained for the simulated collector. The simulated collector has been assumed to terminate in an end overfall, followed by the treatment plant.

The two analysed CSO devices have been considered to be placed close to the downstream collector end (25 m from the end). Their geometrical characteristics have been fixed according to the usual design criteria and to the value of $Q_{treatmax}$.

Following the Italian procedures for the CSO devices design, the overflow discharge process has to start in correspondence of a dilution threshold value of flow rate $Q_{thres} = r\,Q_d$, where r is a dilution coefficient, to be fixed conveniently within the range 3-6. Obviously, the adoption of lower values of r determines an increase of CSO discharges and a consequent decrease of treated volumes. In particular, the extreme values $r = 3$ and $r = 6$ have been considered in the present work for the two side-weirs. The hydraulic and geometrical characteristics of the two simulated CSO devices are reported in Tables 1 and 2.

Table 1 Hydraulic and geometrical characteristics of the low crested CSO device.

	Q_{thres} (m^3/s)	$Q_{treatmax}$ (m^3/s)	Crest height (m)	Weir length (m)
$r = 3$	0.021	0.091	0.06	17.0
$r = 6$	0.042	0.134	0.09	18.0

Table 2 Hydraulic and geometrical characteristics of the high crested CSO device.

	Q_{thres} (m^3/s)	$Q_{treatmax}$ (m^3/s)	Crest height (m)	Weir length (m)	Orifice opening (m)
$r = 3$	0.021	0.091	0.60	5.0	0.06
$r = 6$	0.042	0.134	0.60	5.0	0.09

Experimental data

Experimental hydrographs and pollutographs relative to seven rain events recorded during the period 2000-2001 have been considered in the present work. Among the different measured pollutants, suspended solids (SS) have been used for the simulations.

Table 3 reports some characteristics of the analysed events. As examples of the available data, Figure 2a and Figure 2b report the hydrographs $Q(t)$ and pollutographs $C_{SS}(t)$ of the suspended solids relative to the event n. 2 and the event n. 3 respectively. In particular, Figure 2a shows a last flush (the maximum of the SS concentration occurs after the flow rate peak), while Figure 2b shows the occurrence of a first flush.

Table 3 Characteristics of the simulated events.

Event Number	Event date	Total rain depth (mm)	Hydrogr duration (min)	Max. flow rate (m^3/s)	Max. SS Concentr. (kg/ m^3)	Water volume (m^3)	SS mass (kg)
1	23/6/00	16.4	96	0.551	0.89	765	292
2	8/7/00	7.0	29	0.325	2.96	237	420
3	10/7/00	11.0	69	0.376	1.00	461	133
4	17/3/01	26.2	426	0.281	1.28	1445	196
5	28/3/01	18.6	437	0.263	2.36	985	205
6	10/4/01	8.4	108	0.157	1.20	450	191
7	20/4/01	15.8	368	0.257	1.19	910	340

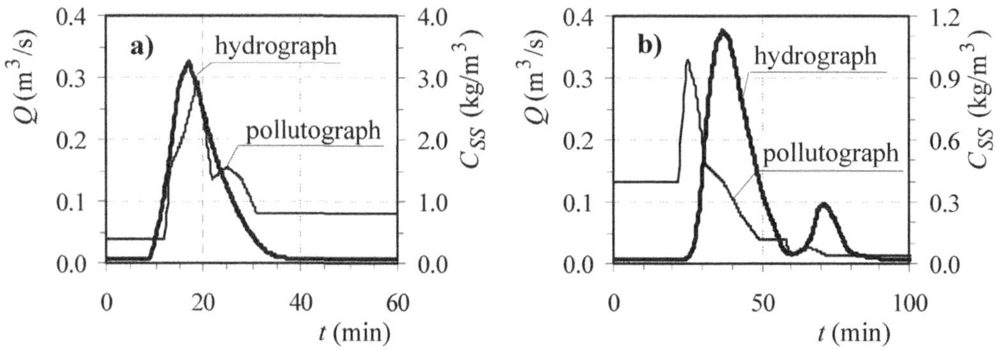

Figure 2 Hydrographs and pollutographs relative to: a) event n. 2; b) event n. 3.

Simulations and results

In order to compare the two CSO devices, the application of the numerical model has been carried out for the seven events separately for each device. Measured flow rates and pollutant concentrations have been considered to enter from the collector upstream end and to propagate downstream to the CSO device. Simulations have allowed for computing the hydrographs and pollutographs of the overflow discharges and of the flows to the treatment plant. As an example, figure 3 reports the entering flow rates, the CSO flow discharges and the flow rates conveyed to the treatment plant for the event n. 4 and $r = 6$.

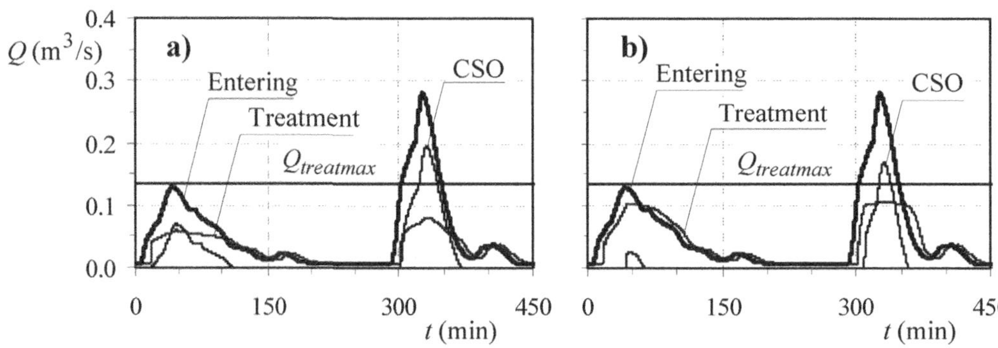

Figure 3 Computed hydrographs for the event n. 4 and $r = 6$: a) low crested CSO device; b) high crested CSO device.

The figure shows that the water volume globally discharged by the high crested CSO device results remarkably lower than the volume discharged by the low crested CSO device. Moreover, although the high crested side-weir increases the treated volume, according to the device design criteria, the flow rates conveyed to the treatment plant never exceed the maximum treatment capacity $Q_{treatmax}$. A similar comparison concerning the suspended solid mass rates Q_{SS} discharged through the CSO devices and conveyed to the treatment plant is reported in figure 4. The figure shows the reduction of the discharged mass and the increase of the treated mass when the high crested CSO device is adopted.

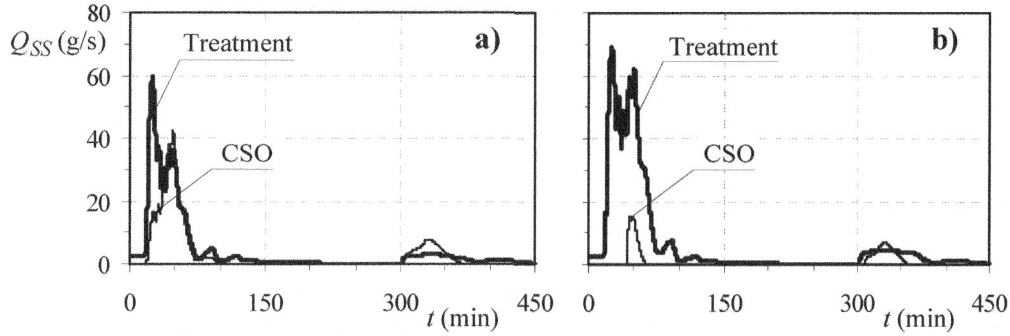

Figure 4 Computed pollutant mass rates relative to the event n. 4 and $r = 6$: a) low crested CSO device; b) high crested CSO device.

Results concerning all the analysed events are summarised in figures 5 and 6. In particular, figure 5 reports the total discharged water volume W for the two devices and for the two values of r. Similarly, in figure 6 the total discharged SS mass P is reported.

Globally, the analysis of the results of all the events has shown that the high crested CSO device provides significant reductions of both discharged volumes (about 44% for $r = 3$ and 50% for $r = 6$) and discharged SS masses (about 45% for $r = 3$ and 51% for $r = 6$), in comparison to the low crested side-weir.

Additional simulations have also shown that results are poorly sensitive to the values of the dispersion coefficient; as expected, in fact, the effects of the advection transport component during wet weather flows in sewer systems result significantly predominant than the effects due to the dispersion component.

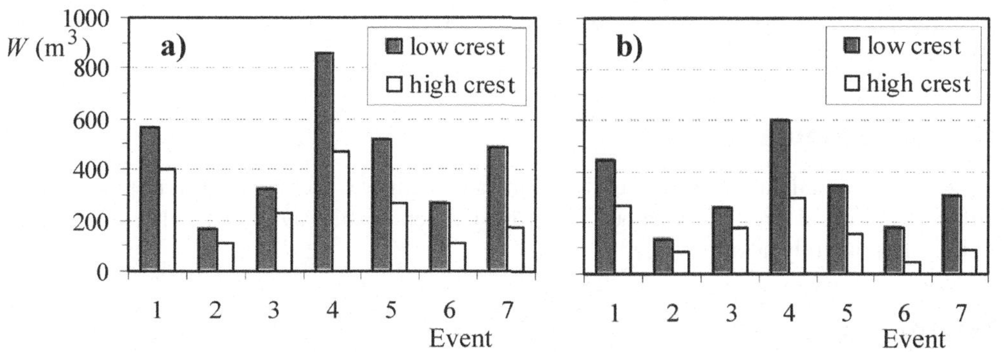

Figure 5 Total discharged water volume W for the two CSO devices: a) $r = 3$; b) $r = 6$.

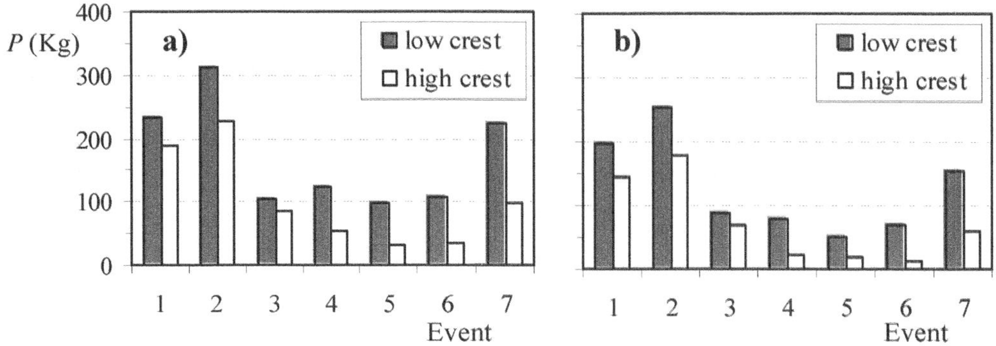

Figure 6 Total discharged SS mass P for the two CSO devices: a) $r=3$; b) $r=6$.

Conclusions

In the last decades, environmental problems have induced scientists and professionals to consider adequate CSO devices able to reduce, during rain events, the sewer overflow discharges into water bodies, both in terms of water volumes and pollutant masses.

This paper shows the results of an analysis on the quantitative and qualitative performances of the high crested side-weirs with bottom orifice. A comparison between the performances of this CSO device and the more common low crested side-weirs CSO device is also presented.

In order to perform the analysis, an accurate physically based 1-D model solving the De Saint Venant equations for the water flow and the advection-dispersion equation for the pollutant propagation has been implemented. Literature well-established values have been used for the side-weir discharge coefficient and for the pollutant dispersion coefficient. Simulations have been run using hydrographs and pollutographs recorded in the experimental urban catchment of Cascina Scala, Italy, during different rain events.

The results of the simulations show that for all the examined events the high crested CSO device can perform a remarkable reduction of discharged volumes and pollutants in comparison to the low crested side-weir. In addition, the high crested CSO device increases the treated volume and allows for a more efficient use of the treatment plant capacity.

Future outlooks could concern the simulation of the performances of the devices by using more complex 2-D models which could allow for taking into account the concentration distribution along the vertical.

Acknowledgements

The experimental data concerning the urban catchment of Cascina Scala were provided by Prof. S. Papiri and Prof. C. Ciaponi from the University of Pavia, Italy. Their help and advices are gratefully acknowledged.

References

Andoh, R.Y.G. and Saul, A.J. (2002) The use of hydrodynamic vortex separators and screening systems to improve water quality. In *Proceedings of 3rd International conference on Sewer, Processes and Networks*, Paris, pp. 219-229.

Armanini, A. (1999) Principi di idraulica fluviale. Editoriale Bios, Cosenza.

Artina, S., Calenda, G., Calomino, F., La Loggia, G., Modica, C., Paoletti, A., Papiri, S., Rasulo, G. and Veltri, P. (1997) Sistemi di Fognatura. Manuale di Progettazione. HOEPLI, Milano.

Bertrand-Krajewski, J.L., Campisano A., Creaco, E. and Modica, C. (2004) Experimental study and modelling of the hydraulic behaviour of a hydrass flushing gate. In *Proceedings of NOVATECH 2004*, Lyon, June 6-10, pp. 557-564.

Brombach, H. (1992) Solids Removal from Combined Sewer Overflows with Vortex Separators. In *Proceedings of NOVATECH 92,* Lyon, November 3-5, pp.447-459.

Campisano, A., Creaco E. and Modica C. (2004) Experimental and numerical analysis of the scouring effects of flushing waves on sediment deposits. Accepted for publication on *Special Issue of Journal of Hydrology.*

Ciaponi, C., Mutti, M. and Papiri, S. (2002) Storm water quality in the Cascina Scala (Pavia, Italy) experimental catchment. In *Proceedings of the 2nd International conference on New trends in Water and Environmental Engineering for Safety and Life: Eco-compatible Solutions for Aquatic Environments*, p. 208.

Ciaponi, C. and Papiri, S. (1994) Bacino di Cascina Scala - Codice IT04. Le misure di pioggia e di portata nei bacini sperimentali urbani in Italia, Centro Studi Deflussi Urbani, Milano, pp. 1-127.

Deletic, A. and Maksimovic, C. (1998) Evaluation of water quality factors in storm runoff from paved areas. *Journal of Environmental Engineering*, ASCE, **124**(9), 869-879.

Elder, J.W. (1959) The dispersion of marked fluid in turbulent shear flow. *Journal of Fluid Mechanics*, (5), 544-560.

Field, R. (1972) The Swirl Concentrator as a CSO Regulation Facility. USEPA-R2-72-008.

Field, R. (1990) Combined sewer overflows: control and treatment. Control and Treatment of Combined Sewer Overflows, P.E. Moffa Edition, Environmental Engineering Series, Van Nostrand Reinhold, New York, NY.

Garcia-Navarro, P., Alcrudo, F. and Saviron, J.M. (1992) 1-D Open-Channel Flow Simulation Using TVD-McCormack Scheme. *Journal of Hydraulic Engineering*, ASCE, **118**(10), 1359-1372.

Garcia-Navarro, P. and Saviron, J.M. (1992) McCormack's method for the numerical simulation of one-dimensional discontinuous unsteady open channel flow. *Journal of Hydraulic Research*, ASCE, **30**(1), 95-104.

Garsdal, H., Mark, O., Dørge, J. and Jepsen, S.E. (1995) Mousetrap: Modelling of water quality processes and interaction of sediments and pollutants in sewers. *Water Science and Technology*, **31**(7), 33-41.

Ghetti, A. (1984) Idraulica. Libreria Cortina, Padova.

Gray, S.R. and Becker, N.S.C. (2002) Contaminant flows in urban residential water systems. *Urban Water*, Elsevier Science, **4**(4), 331-346.

Mark, O., Hernebring, C. and Magnusson, P. (1998) Optimisation and control of the inflow to a wastewater treatment plant using integrated modelling tools. *Water Science and Technology*, **37**(1), 347-354.

Ogata, A. and Banks, R.B. (1961) A solution of the differential equation of longitudinal dispersion in porous media. U.S. Geol. Surv., *Prof. Pap. No. 411-A*.

Payne, J.A., Moys, G.D., Hutchings, C.J. and Henderson, R.J. (1990) Development, calibration and further data requirements of the sewer flow quality model Mosquito. *Water Science and Technology*, **22**(10/11), 103-109.

Papiri, S. (2000) Gli scaricatori di piena nelle fognature miste alla luce dei risultati di una simulazione continua quali-quantitativa delle acque meteoriche nel bacino urbano sperimentale di Cascina Scala (Pavia). In *Atti della II Conferenza Nazionale sul Drenaggio Urbano*, Palermo, pp. 49-60.

Saul, A.J. and Delo, E.A. (1982). Laboratory tests on a sewer overflow chamber with unsteady flow. In *Proceedings of the First International Seminar on Urban Drainage System*, Southampton.

Schlütter, F. and Mark, O. (2002) Dynamic modelling of pollutants from CSO's. In *Proceedings of 3rd International conference on Sewer, Processes and Networks*, Paris, pp. 211-218.

Smisson, B. (1967) Design, Construction and Performance of Vortex Overflows. In *Proceedings of thee Symposium on Storm Sewage Overflows*, Institute of Civil Engineers. London, pp. 99-110.

Xanthopoulos, C. and Hahn, H.H. (1993) Anthropogenic pollutants wash-off from street surfaces. In *Proceeding of the 6th International Conference on Urban Storm Drainage*, Niagara Falls, pp. 417-422.

Spatial evolution of the characteristics of urban wastewater in combined sewers

M. Kafi-Benyahia*, M.G. Gromaire* and G. Chebbo**

*CEREVE, ENPC, 6 et 8 av. Blaise Pascal, 77455 Marne La Vallée Cedex, France, (Tél:33.(0)1.64.15.36.25, Fax : 33.(0)1.64.15.37.64), (e-mail : benyahia@cereve.enpc.fr, gromaire@cereve.enpc.fr)
**Faculté de Génie, Université Libanaise, Route de l'aéroport, Beyrouth, Lebanon, (e-mail: chebbo@cereve.enpc.fr

Abstract An experimental on-site observatory of urban pollutant loads in combined sewers was created in the centre of Paris to quantify and characterise the dry and wet weather flow depending on spatial scale. Dry weather flow was monitored for seven days at the outlet of the six nested urban catchments and was analysed for suspended solids, organic matter, nitrogen and heavy metals. Results show a relative stability of pollutant loads of SS, organic matter and nitrogen, and important variations of heavy metals loads from one site to another. Most part of the pollutants were transported in particulate form. At last, an increase for organic matter and nitrogen loads appeared between the sampling sites situated at the upstream and the middle of the catchment area of study. Results obtained by dry weather, will be used as reference for the characterisation of wet weather flow pollution.

Keywords combined sewer, experimental catchments, heavy metals, organic matter, pollutant loads, spatial scales, wastewater.

Introduction

Several Several studies led since the 70's have underlined the importance of the urban wet weather flow pollution and their impact on the receiving waters (Torno, 1984; Bachoc, 1992; Chebbo, 1992; Saget, 1994, Pisano; 1996).
The research led especially by Gromaire *et al* (2001) and Chebbo *et al* (2001) on "Le Marais" experimental catchment, in Paris (42 ha), between 1995 and 2000, has shown that the sewer network is not only a system of transport but also physico-chemical reactor which conditions by its characteristics the quality of urban waters. This results indicated an evolution of the characteristics (loads and nature) of wet weather flow between the input and the output of the sewer network (increase in concentrations of SS and organic matter, decrease in concentrations of heavy metals, and increase in percentage of particle bound pollution); and shown that 40 to 80% of SS comes from in-sewer pollutants stocks.
Taking into account these results, CEREVE in partnership with the municipality of Paris, SIAAP and the Seine Normandy Water Agencies, created in Paris in 2002 an experimental on-site observatory of urban pollutant loads in combined sewer systems "OPUR"(Observatory of Urban Pollutants in Paris). This on-site observatory will enable to study developments in water quality between the upstream and the downstream of a large catchment area, main sources of pollution and dominant mechanisms in relation to the spatial scale, and pollution transfer and transformation processes on a more local scale.

 Sewer Networks and Processes within Urban Water Systems edited by J.-L. Bertrand-Krajewski, M. Almeida, J. Matos & S. Abdul-Talib. ISBN: 1 84339 506 1.

This article describes the experimental methods used for this research and the first results obtained concerning the quality of wastewater during dry weather at the exit of each urban catchment. These results aim at the study of variability of characteristics of wastewater for a same type of agglomeration depending on the size from the catchment area, the duration of transfer in sewer and the land use of urban surface. They supply an interesting elements for the conception and the dimensioning of wastewater treatment plant. Wastewater constitute furthermore one of sources of pollution of combined wet weather flow, which the knowledge is necessary within the framework of OPUR research program, for the spatial evolution of wet weather flow.

Experimental methods

On-site observatory

The six experimental sites of OPUR are situated in central Paris, along the main drain constituted by the Clichy trunk sewer (Figure 1). The corresponding urban catchments expand-on a surface from 40 ha to 2320 ha, with sewer length from 1 km to 13 km and a dense population (from 12 000 inhabitants to 620 000 inhabitants). The sewerage system is combined and completely man entry. The details of characteristics of the six OPUR catchments are given in Table 1:

Table 1: characteristics of the six OPUR experimental sites

Urban catchment	Area (ha)	Hydraulic length (m)	Populations
Marais	42	1000	12 400
Sébastopol	112	2500	31 000
Quais	438	5000	99 400
Clichy centre	943	9000	211 400
Coteaux aval	1098	9000	349 430
Clichy aval	2311	12500	619 980

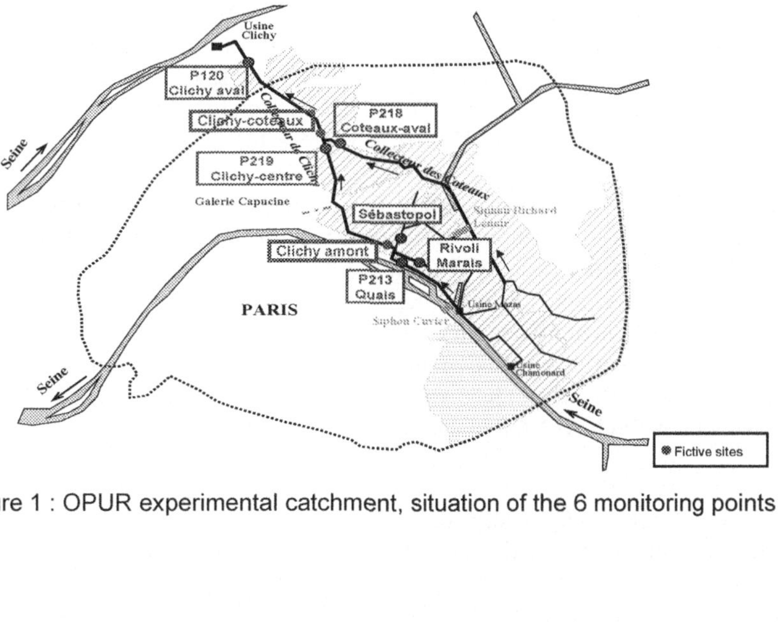

Figure 1 : OPUR experimental catchment, situation of the 6 monitoring points

Several connections exist between the OPUR area and the neighbouring urban catchments. A most of these connections are actives for important rain events. By dry weather, OPUR receive waters from the Rive Gauche of Paris (400 l/s[1]) through Siphon Cuvier.
Moreover in consequence of renovation works realised on Asnières collector's, 587[2] l/s was given by Capucine Gallery to Clichy collector's during our dry weather campaign. These volumes were subtracted from the volumes measured at Quais, Clichy centre and Clichy aval before to calculate the flow and loads per inhabitants.
On account of the cleansing works realised at Rivoli sewer's in the Marais catchement, the results of this site will be replaced by those obtained by Gromaire in 1998.

Monitoring equipment

The OPUR experimental catchments were provided with equipment allowing to characterise and quantify flow and pollutant loads at the catchment outlet during dry and wet weather.
At the outlet of each catchment, flow was monitored by an flow meter and recorded every 2 to 6 minutes. Water velocities were measured on three to four horizontal lines in the cross section with a time of flight ultrasonic system. The water level was measured both with an ultrasonic and pressure sensor at Marais and with one or two pressure sensors at the other sites. Samples were taken at each site with two automatic vaciometric samplers which were driven by the flow meter and contained 24 bottles each. They were used to constitute daily mean sample or pollutographs. 15 rain gauges belonging most of them to municipality of Paris, cover the studied area.

Analyses

The following pollutant parameters were studied : suspended solids "SS", organic matter (volatile suspended solids "VSS", chemical oxygen demand "COD", biochemical oxygen demand "BOD_5", particulate organic carbon "POC", dissolved organic carbon "DOC"), nitrogen (ammonia "NH4+", total Kjeldahl nitrogen "TKN"), heavy metals (Cd, Cu, Pb, Zn), settling velocities, pH and conductivity.
All parameters were analysed both on the total and on the dissolved fractions. The separation between the dissolved and the solid fraction for SS, COD, DBO_5, DOC, NH4+ and TKN was made by filtration under vacuum on glass fibre filters (Whatman GF/F). Before use, the filters have been washed and precombusted at 525°C during 1h. SS were measured by weighting the retained mass after 2h drying at 105°C. The same filters were further calcined at 525°C and weighed to measure VSS. Dissolved and total BOD_5 were measured following the French AFNOR standards (NF T90-103) whereas for the dissolved and total COD the HACH micro-method was used.
The analysis of DOC were realised using the French AFNOR standards (NF EN 1484). However for the POC, measurement were accomplished on the filters used for SS measurement. After 1h drying at 105°C, these filters were weighted and were placed on aluminium slabs in glass petri dishes. They were then acidified with hydrochloric acid and were analysed by COT-meter when they were dry.
For NH4+, the analysis were realised on the dissolved fraction following the French AFNOR standards (NF T90-015). The TKN was measured both on the dissolved and the total fractions using the French AFNOR standards (NF EN 25663).

---

[1] Mean flow of Siphon Cuvier by dry weather.
[2] This value was estimated from the modelling of network functioning.

For dissolved metals, 50-100ml of the total samples were filtered at 0.45µm. After that, the filtrate was acidified at pH 1 with 2ml of the concentrated nitric acid and stored at 4°C before analysis. For the total metals, 100ml of bulk samples were digested with concentrated perchloric acid (2ml) and concentrated nitric acid (18ml). The liquid was evaporated to dryness in a sand bath at 110°C. The final residue was dissolved using 30ml of nitric acid. After 15 min-centrifugation (at 3500rpm), the supernatant was stored at 4°C before analysis. Finally, dissolved and total metals were determined using ICP-MS.

Results and discussion

Characterisation of waste water flow

Dry weather flow has been characterised at the outlet of the urban catchments for 4 to 7 days of the week, between October-November 2002 and February-Marsh 2003.

Flow

The total daily flow on the six catchments is illustrated in Figure 2 and varies from 420 l/head/day at Quais to 820 l/head/day at Clichy centre. These values were very high compared with those given by Saget and Chebbo (1996) [100 to 300 l/head/day], but they were comparable with those obtained by Philippe *and al* (1987) [398 to 485 l/head/day], except for Clichy centre. High variations from one dry day to another day was observed in particular for the Marais, Sébastopol, Quais and Clichy centre.

This high production can be explained at Paris by the importance of street washing, business activities (restaurant, trade, offices..) and the probable contribution of parasite waters. However for Clichy centre, it can be explained by more producer activities that they must be identified with an analysis of the land use of urban surface, or it can be linked to contribution of water originated from the functioning of some connections.

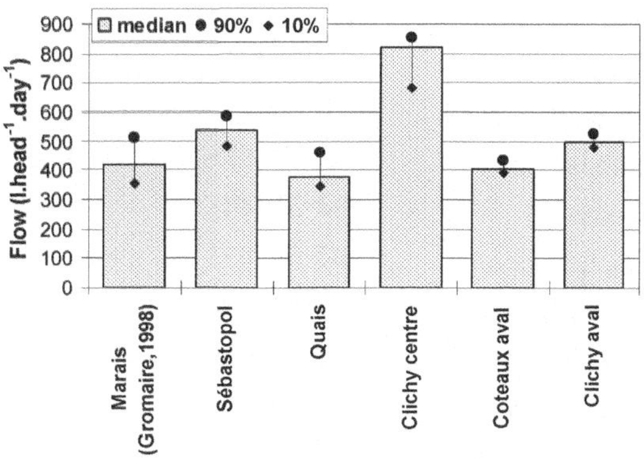

Figure 2 : spatial evolution of the total daily flow on the OPUR sites

Pollutant loads

Suspended solids, oxygen consuming matter and nitrogen

Table 2 and Table 3 give the pollutant loads (SS, VSS, COD, BOD_5, TOC, TKN and NH4+) of wastewater measured at the outlet of the six catchments. For a best study of the spatial evolution, two fictive sites were added: Clichy amont (sum of "Marais, Sébastopol, and Quais) and Clichy-Coteaux (sum of "Clichy centre and Coteaux aval). These fictive sites were calculated in order to be compared respectively with Clichy centre and Clichy aval.

Table 2: pollutant loads of SS, and consuming matter of wastewater

	SS (g/head/day)	VSS (g/head/day)	COD (g/head/day)	BOD_5 (g/head/day)
Marais (Gromaire,98)	58-75 **[65]**	50-64 **[57]**	135-181 **[159]**	62-85 **[76]**
Sébastopol	78-117 **[91]**	66-108 **[83]**	158-234 **[196]**	31-114 **[75]**
Quais	61-88 **[68]**	52-73 **[57]**	130-154 **[142]**	57-74 **[64]**
Clichy amont	*69-97 **[75]***	*59-86 **[66]***	*157-166 **[158]***	*48-73 **[60]***
Clichy centre	122-159 **[129]**	111-137 **[117]**	227-318 **[248]**	113-149 **[139]**
Coteaux aval	85-98 **[87]**	73-84 **[74]**	161-208 **[182]**	60-71 **[67]**
Clichy-Coteaux	*103-116 **[105]***	*89-100 **[93]***	*204-225 **[218]***	*50-98 **[90]***
Clichy aval	90-103 **[100]**	78-92 **[88]**	158-185 **[173]**	80-93 **[87]**
Philippe and al (87)	31 to 71	–	60 to 140	29 to 65
People equivalent (a)	90	–	57*	
People equivalent (b)			120 to 135	57 to 70

Notes: 10% - 90% **[median]**, Clichy amont and Clichy-Coteaux are respectively the sum of "Marais, Sébastopol, Quais" and "Clichy centre, Coteaux aval", (a) French *pe*, (b) other European *pe*. (quoted by F. Fievet, J. Lesavre; 98), *value of organic matter: $OM=(2*DBO_5+DCO)/3$.

Table 3: pollutant loads of TOC, and nitrogen of wastewater at the outlet of the six catchments

	TOC (g/head/day)	TKN (g/head/day)	NH4+ (g/head/day)
Marais (Gromaire,98)	–	–	4.2-4.9 **[5]**
Sébastopol	47-67 **[57]**	18-20 **[19]**	10-12 **[11]**
Quais	38-51 **[44]**	12-17 **[14]**	7.1-9.6 **[7.3]**
Clichy amont	*45-54 **[47]***	*14-17 **[16]***	*7.9-10.1 **[8]***
Clichy centre	67-94 **[79]**	21-25 **[22]**	14.7-18 **[14.9]**
Coteaux aval	48-57 **[53]**	15.4-16 **[15.7]**	10-11.4 **[10.3]**
Clichy-Coteaux	*60-69 **[61]***	*17-19 **[18]***	*11.9-13.6 **[12.3]***
Clichy aval	51-59 **[55]**	16.7-18.2 **[17.2]**	10.6-12.5 **[12]**
People equivalent (a)	–	15	–

The daily mean loads of the different pollutants were relatively stables between Marais, Sébastopol, Quais and Coteaux aval (the variance[3] is 28% for SS, 16% for BOD_5 and 22% for TOC). However, it was high at Clichy centre and Clichy aval (248 to 173 g/head/day for COD and 139 to 87 for BOD_5). The values measured at Clichy centre were forever much higher than the literature. The pollutant load of SS was in the upper range of those obtained by Philippe *and al* (1987) at Marais and Quais, but it was usually high at the other sites. However, the COD and BOD_5 loads were high in comparison with the values of people equivalent and those kept by Philippe *and al* (1987). Values of TKN were comparable with the values of people equivalent except for Clichy centre.

[3] variance = (maximal value – minimal value)/ maximal value

This variation is greatly linked to the importance of professional activities on urban catchments. It can be linked however at Clichy centre to the difference of the land use between intermediate urban catchments or to the contribution of waters which were not identified. A variability appeared from one dry day to another.
Finally, an increase in pollutant loads was observed between Clichy amont (sum of "Marais, Sébastopol, Quais") and Clichy centre, however the pollutant loads of "Clichy-Coteaux" (sum of "Clichy centre and Coteaux aval") were comparable with those of Clichy aval (except for COD where the pollutant loads were relatively high for the $\sum$(Cc, Ca) with 26% of variance).

Heavy metals

The range of Cd, Cu, Pb and Zn loads measured at the OPUR sites are illustrated by Figure 3.

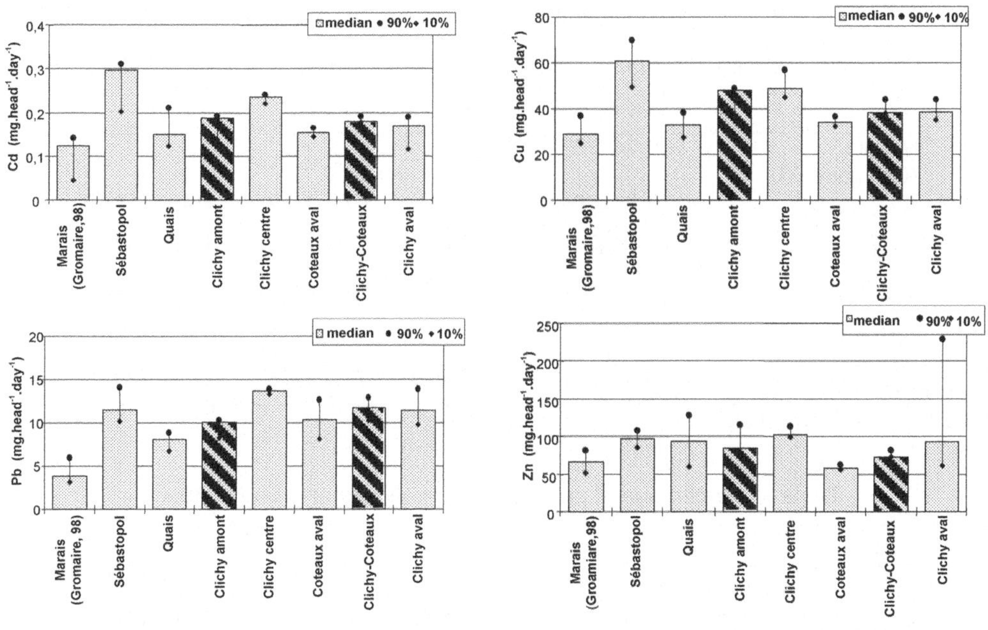

Figure 3 : Spatial evolution of daily mean load of heavy metals at the OPUR sites

For heavy metals, unlike the other parameters of pollution, loads varied considerably depending on the sampling sites. The highest values were measured at the outlet of the Sébastopol (for cadmium and copper) and at Clichy centre (for lead and zinc). The range of pollutant loads of heavy metals obtained at Marais (by Gromaire,1998) and at Coteaux aval were lower to those measured by Philippe *and al* (1987) [0.18-4.7 mg/head/l for Cd, 8.6-37 mg/head/l for Pb, 99.5-253 mg/head/l for Zn). However, those measured at Quais, Sébastopol, Clichy centre and Clichy aval were relatively comparable in comparison with literature.
Heavy metals loads varied significantly from one day to another day. This variability can be explained by punctual pollution linked to industrial activities. Different behaviours between the Cd, Cu, Pb and Zn loads was observed and showed that heavy metals have different sources. One can notice from comparison between the sum of "Marais, Sébastopol, Quais"

and Clichy centre and the sum of "Clichy centre, Coteaux aval" with Clichy aval that the pollutant loads of heavy metals were stables along the Clichy collector's.

Proportion pollution linked to particle

67 to 82% of the oxygen consuming matter and 40 to 93 % of heavy metals were transported in particulate form, whereas, the nitrogen proportion varied between 16 to 22% (Table 4 and Table 5). This a normal phenomenon because TNK is mainly presents in wastewater in NH4+ form.

Distribution between dissolved and particle bound pollution of oxygen consuming matter was relatively stable from a sampling site to another. It was relatively stable too for heavy metals expect for Sébastopol for Cd and Zn. The percentages of copper and lead were to a large extent in particulate form in comparison with cadmium and zinc.

Comparison with the percentages measured at OPUR sites and those obtained at Marais in 1998 by Gromaire, showed that they were in the same rang for the distribution of oxygen consuming matter but they were comparable for Cu and Zn.

Table 4: percentage of particle bound pollution of oxygen consuming matter and nitrogen

	%DCO	%BOD5	%COP	%TKN
Marais (Gromaire,98)	66-74 **[70]**	63-70 **[67]**	–	–
Sébastopol	70-80 **[76]**	68-81 **[78]**	67-80 **[72]**	15-20 **[16]**
Quais	72-74 **[74]**	68-75 **[72]**	70-74 **[72]**	17-21 **[20]**
Clichy centre	79-87 **[82]**	73-83 **[80]**	75-81 **[81]**	16-25 **[22]**
Coteaux aval	78-84 **[81]**	67-76 **[72]**	76-80 **[79]**	17-43 **[18]**
Clichy aval	86-98 **[81]**	72-78 **[75]**	77-79 **[78]**	17-24 **[20]**

Note: 10% - 90% **[median]**

Table 5 : percentage of particle bound pollution of heavy metals

	%Cd	%Cu	%Pb	%Zn
Marais (Gromaire,98)	100*	89-94 **[93]**	46-95 **[61]**	65-96 **[71]**
Sébastopol	9-75 **[43]**	73-83 **[78]**	80-85 **[82]**	40-69 **[47]**
Quais	61-80 **[75]**	76-91 **[82]**	78-90 **[85]**	61-78 **[68]**
Clichy centre	57-74 **[69]**	87-90 **[89]**	73-81 **[76]**	45-70 **[64]**
Coteaux aval	66-73 **[68]**	84-90 **[87]**	79-90 **[89]**	56-76 **[68]**
Clichy aval	59-73 **[66]**	81-91 **[89]**	62-86 **[84]**	60-68 **[62]**

Note: 10% - 90% **[median]**, *Cd dissolved concentrations were lower than detection threshold

Nature of the particle

Figure 4 shows COP and TKN contents of suspended solids from wastewater measured at the exit of the six urban catchments.

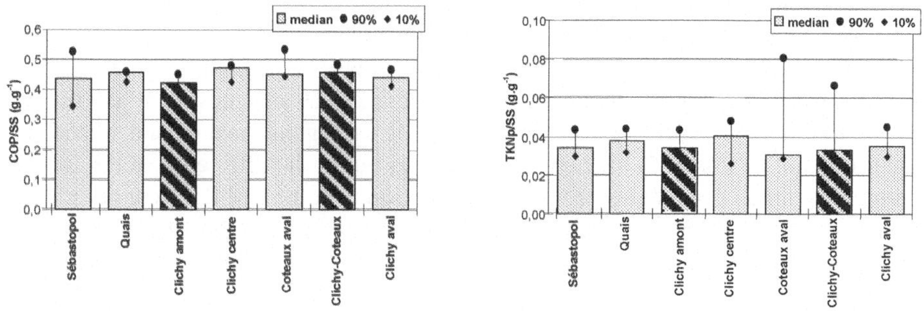

Figure 4: spatial evolution of COP and TKN contents of suspended solids

COP contents varied from 0.44 to 0.47 and TKN contents varied from 0.038 to 0.041. Variation between the sampling sites was stable. A high variability from one dry day to another was however observed. One can notice that the COP and TKN contents were stable along the Clichy trunk sewer.

As illustrated by Figure 5, the nature of heavy metals content of particles differd from one site to another. The highest values of Cd, Cu, Pb and Zn contents of particles were measured at Marais for Cd, at Sébastopol for Cu, at Sébastopol and Coteaux aval for Pb and at Quais for Zn. A very significant variation from one day to another day was established and might be explained by the industrial contributions which are variables in time.

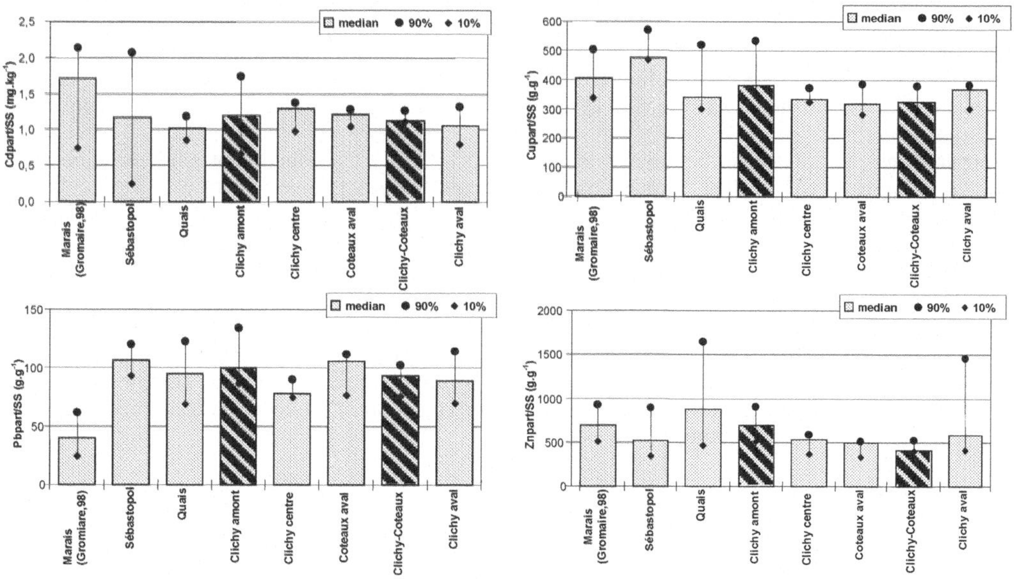

Figure 5: spatial evolution of heavy metals contents of suspended solids

Conclusions

This research has consisted of characterisation of wastewater quality in combined sewer at various spatial scales. Results obtained show, a relative stability of pollutant loads between,

Marais, Sébastopol, Quais and Coteaux aval, a high values at Clichy centre, and a different behaviour of heavy metals loads from one site to another. They showed also that pollutant were mainly transported in particulate form. An increase depending on the spatial scale appeared between the three urban catchments situated at the upstream of the Clichy collector's and Clichy centre for the pollutant loads of organic matter and nitrogen. This increase might not be linked to the network processes but to the difference of the land use of the intermediate urban catchments or to the contribution of waters which were not identified. Wastewater quality will be used as reference for the characterisation of wet weather flow pollution. So a study of water quality in combined sewers by wet weather is in progress. The results obtained for dry and wet weather should enable the development of a methodology allowing the competent bodies to choose the control strategy that is best adapted to the characteristics of the storm water.

References

Bachoc, A. (1992). Le transfert des solides en réseaux d'assainissement unitaires. *Ph.D,* Institut Nationale Polytechnique de Toulouse, France, 281p. + appendices.

Chebbo, G. (1992). Solides des rejets pluviaux urbains, caractérisation et traitabilité. *Ph.D,* Ecole Nationale des Ponts et Chaussées, 413 p. + appendices.

Chebbo, G., Mouchel, JM., Saget, A. and Gousailles, M. (1995). La pollution des rejets urbains de temps de pluie: flux, nature et impacts. T.S.M., 11, 796-806.

Chebbo, G, Gromaire-Metz, MC, Ahyerre, M., Garnaud, S. (2001). Production and transport of urban wet weather pollution in combined sewer systems: the "Marais" experimental urban catchment in Paris. Urban Water, vol.3, n°1-2, p. 3-15.

Degrémont (1989). Mémento technique de l'eau. Prais, Lavoisier: Technique et Documentation, 1459p.

Gromaire-Metz, MC. (1998). La pollution des eaux pluviales urbaines en réseau d'assainissement unitaire-Caractéristiques et origines. *Ph.D,* Ecole Nationale des Ponts et Chaussées, 506 p. + appendices.

Gromaire-Metz, MC., Garnaud, S., SAAD, M., Chebbo, G. (2001). Contribution of different sources to the pollution of wet weather flows in combined sewers. Water Research, vol.35, n°2, p.521-533.

Philippe, JP., and Ranchet, J. (1987). Pollution des eaux de ruissellement pluvial en zone urbaine. Synthèse des mesures sur dix bassins versants en région parisienne. Research Report LPC n° 142. Ministère de l'Equipement du Logement et des Transports, Laboratoire Central des Ponts et Chaussées, 70 p.

Pisano W.C. (1996). Summary: United states "sewer solids" settling characterisation methods, results, uses and perspective. Water Science and technology, vol 33, n°9, p 109-115.

Saget, A. (1994). Base de données sur la qualité des rejets urbains de temps de pluie : distribution de la pollution rejetée, dimensions des ouvrages d'interception. *Ph.D,* Ecole Nationale des Ponts et Chaussées, 227 p. + appendices.

Needs and influence of calibration and validation data sets in stormwater quality models of various levels of complexity

M. Mourad *, J.-L. Bertrand-Krajewski *, G. Chebbo **

* Laboratoire URGC Hydrologie Urbaine, INSA de Lyon, 34 avenue des Arts, 69621 Villeurbanne cedex, France.
** CEREVE, ENPC, 6-8 av. Blaise Pascal, Cité Descartes Champs sur Marne, 77455 Marne la Vallée Cedex 2, France.

Abstract Stormwater quality modelling has become a fundamental issue in urban hydrology. Practitioners are led to work with Stormwater Quality Models (SQMs), the operational use of which remains a difficult task despite the progress achieved during the last two decades. A method to assess the variability in calibration and validation results of various SQMs with different levels of complexity due to the variability in calibration data sets is proposed. Preliminary results for site mean concentration (SMC) and Event Mean Concentration (EMC) models used to evaluate SS and COD concentrations in "Le Marais" catchment show that a minimum of 15 to 20 rainfall events is needed to calibrate them in order to get acceptable results compared to what is achievable by using more rainfall events. The size of validation data sets appears less critical for EMC models: if 20 to 40 events are used for validation, the validation quality remains rather constant for SS models and showed little variations for COD models. These conclusions strongly depend on the total available data sets and on the fraction used for calibration.

Keywords Stormwater, modelling, calibration, validation, uncertainty.

Introduction

During the last two decades, research on stormwater quality models (SQM) has been very active and led to the development of numerous models, with a wide variety of concepts and approaches. Satisfactory results given by some models have been reported (e.g. Zug, 1998), but always with limitations and precautions to be considered by model users. A trend to develop more complex models, based on increasingly detailed descriptions of involved processes and incorporating some hardly measurable or identifiable variables can be observed (Ashley *et al.*, 1999). At least in France, practitioners in cities, consulting companies and environment agencies are rarely using such complex models which were usually developed by researchers before being implemented in commercial softwares (Gromaire *et al.* 2002). This could be explained by the great expense and effort needed for the application of complex models. Simpler approaches such as the simple method based on site mean concentration (SMC) (Schueler, 1987) and event mean concentration (EMC) (Driver and Tasker, 1990; Saget, 1994) are believed to be suitable and sufficient for some cases where stormwater concentrations and loads need to be evaluated. The multitude of modelling approaches leaves practitioners helpless to choose the most suitable model for a

 Sewer Networks and Processes within Urban Water Systems edited by J.-L. Bertrand-Krajewski, M. Almeida, J. Matos & S. Abdul-Talib. ISBN: 1 84339 506 1.

specific case study and under specific conditions. In most of the cases, rather than a choice problem, it is more likely an optimisation problem. The user should find an appropriate compromise between i) the model complexity, ii) the level of available or attainable knowledge (number and characteristics of measured events to be used for calibration and validation), iii) the objectives of the study and iv) the acceptable level of uncertainty in modelling results (complex models are assumed to provide greater accuracy than simpler models).

A French national research programme has been launched in 2001, funded by the RGCU (Research Network for Civil and Urban Engineering). Its main objectives concern i) the calibration and validation methods applied to SQMs, ii) the comparison of models outputs and uses, and iii) the analysis of operational needs and use of SQMs by practitioners who use models for design and simulations and not for scientific knowledge and purposes. From an operational point of view, one of the questions to be analysed could be formulated as follows: "compared to simple stormwater quality models, or even compared to hydrology-hydraulics models which do not account for pollutant loads, do more complex stormwater quality models provide results which lead to significant differences in terms of operational objectives, like evaluation of sewer systems efficiency, design of overflow structures and of stormwater treatment facilities?".

In order to answer the above question, four typical operational case studies have been chosen : i) evaluation of the pollutant interception efficiency of a sewer system, ii) evaluation of pollutant discharges at a CSO structure, iii) design of a stormwater storage and settling tank, and iv) design of an on-line stormwater settling facility. Various calibrated and validated SQMs with different levels of complexity will be applied, and the results in terms of design and sizing will be analysed. However, before to carry out simulations, it appeared necessary to analyse the needs and the influence of calibration and validation data sets on modelling results and modelling reliability. This first phase is described in this paper, which presents successively : i) the various types of SQMs used, ii) the calibration and validation methods, iii) the experimental catchments and data sets, iv) the results and their discussion.

Types of SQMs used

Three types of SQMs used by practitioners have been distinguished, with increasing levels of complexity:

(1) Site Mean Concentration (SMC) models: the SMC approach consists to attribute to a given catchment a constant concentration for each pollutant. This concentration is commonly calculated as a runoff volume weighted average of the event mean concentrations measured during a given period of time. Other methods for unmonitored sites were developed using regression techniques by Driver and Tasker (1990) from the NURP US national database. Hoos and Sisolak (1993) also proposed an extension of regression models by adjusting predeveloped equations using local data. However, such equations cannot be applied on catchments with different physical and climatic characteristics.

(2) Event Mean Concentration (EMC) models: the EMC approach consists to attribute constant pollutant concentrations to each rainfall event. This concentration is usually calculated as a function of potential explanatory variables such as physical and land-use characteristics of the catchment and rainfall event characteristics. As it is known that no single variable can explain the main part of the EMC value, a multivariable approach is theoretically appropriate. Multiple regression equations are widely used

and, in most of the cases, the models are developed and calibrated with local data sets. Driver and Tasker (1990) also developed EMC regression models for unmonitored catchments.

(3) Build-up/Washoff Models (BWM): Build-up/washoff models aim to simulate assumed governing processes that control runoff quality. These models account for the build-up of pollutants on the catchment and in sewers during dry weather periods and the washoff processes during rainfall events. Available commercial softwares and research models incorporate in-sewer processes such as sediment deposition/erosion, transport and biochemical reactions.

According to Gromaire *et al.* (2002), French practitioners use SMC, EMC and BW models respectively for 43 %, 7 % and 50 % of their projects. Among many existing SQMs belonging to the above three categories, some have been selected. The selection criterion was their implementation in commercial softwares used by French practitioners, like Canoe (Insa/Sogreah, 1999), Mousetrap (DHI, 2000) or InfoWorks from Wallingford Software (Bouteligier *et al.*, 2002). The selected models are described in Table 1.

Table 1 SQMs used in the three categories.

SMC models	$SMC_N = \left(\sum_{i=1}^{N} EMC_i \cdot Vr_i\right) \cdot \left(\sum_{i=1}^{N} Vr_i\right)^{-1}$
EMC models	
Servat (1984)	$EL = K \cdot ADWP^a \cdot I_{max\,5}^b \cdot Vr^c$ $EMC = EL \cdot Vr^{-1} = K \cdot ADWP^a \cdot I_{max\,5}^b \cdot Vr^{c-1}$
Driver and Tasker (1990)	$EL = K \cdot TP^a \cdot D^b$ $EMC = EL \cdot Vr^{-1}$
Cedre (Insa/Sogreah 1999)	$EMC = K \cdot ADWP^a \cdot TP^b \cdot I_{max\,5}^c$
BWMs	
Exponential build-up	$\frac{dM}{dt} = Accu - Disp \cdot M$
Washoff model in SWMM	$\frac{dMa(t)}{dt} = -Ke \cdot I(t) \cdot Ma(t)$
Washoff in Mouse software	$V_{sr}(t) = D_r \cdot \left(\frac{I(t)}{i_d}\right)^{exp} \cdot L \cdot W \cdot (1-\varepsilon) \cdot A_s$
Sediment transport Ackers (1991)	$C_v^* = j \cdot \left(\frac{W_e \cdot R}{A}\right)^{\alpha} \cdot \left(\frac{d_{50}}{R}\right)^{\beta} \cdot \lambda_c^{\gamma} \cdot \left(\frac{\lvert u \rvert}{\sqrt{g \cdot (s-1) \cdot R}} - K \cdot \lambda_c^{\delta} \cdot \left(\frac{d_{50}}{R}\right)^{\varepsilon}\right)^m$

Where *EL* is the event total load; *ADWP* is the antecedent dry weather period (days); I_{max5} is the maximum rainfall intensity on 5 minutes time step (mm/h); *Vr* is the total runoff volume (m^3); *TP* is the total precipitation; *D* is the event duration (hr); *K*, *a*, *b*, *c* are parameters to be adjusted; *M* is the accumulated pollutant load at time *t* (kg); *Accu* is the build-up rate (kg/day); *Disp* is the dispersion rate (day^{-1}); *Ma* is the available load at time t

(Kg); *I(t)* the rainfall intensity at time t (mm/hr); $V_{sr}(t)$ is the volume of eroded sediments (m^3/h); D_r is the detachment rate (m/h); i_d is a reference rainfall intensity (25.4 mm/h); L, W are the length and width of the catchment respectively (m); ε is the sediment porosity; A_s is the surface covered by sediments (m^2). C_v^* is the non-dimensional carrying capacity; R is the hydraulic radius (m); d_{50} is the median particle size (m); s is the specific density; W_e is the effective bed width (m); λ_c is the composite friction factor; g is the gravity (m/s^2); A, α, β, γ, δ, ε, m, n, j, K are coefficients related to the dimensionless grain size D_{gr} with $D_{gr} = d \cdot \left(g \cdot (s-1) \cdot \nu^{-2}\right)^{\frac{1}{3}}$; ν is the kinematic viscosity (m^2/s).

Calibration and validation of SQMs

Model calibration consists to adjust model parameters to better match the observed field output values. A wide variety of calibration methods do exist, ranging from simple to very complex ones. For BWMs, the calibration methods usually applied are rather simple from a mathematical point of view: manual calibration with graphical visual comparisons, use of mathematical metrics (e.g. Calabarò and Maglionico, 2002), or automatic calibration with direct search algorithms (e.g. Zug *et al.*, 1997) or simple random search methods (e.g. Schlütter, 1999). Despite good calibration results, the calibrated model may not provide reliable predictions. Model validation must be performed. It demonstrates the credibility and tests the predictive ability of the developed model. Validation methods can vary for different modelling approaches, but external validation is known to be the most pertinent. This method consists to check the model by using an independent set of field data not previously used in calibration. Good results indicate that the model is "good" and can be used; bad results indicate that either calibration or the model structure needs to be reviewed. Theoretically, the more complex the model, the larger the number of calibration data is required to account as much as possible all potential situations. In practice, the amount of data used for calibration is limited. Consequently, their characteristics are usually not representative of all potential events or situations. This problem can affect considerably the results of the model. Research studies on model calibration are focusing on optimisation methods (e.g. Kanso *et al.* 2002). Sensitivity of stormwater quality models to the number and characteristics of the events used for their calibration have not yet been properly explored. Nevertheless, this question has already been discussed for the calibration of a hydrological rainfall-runoff model (Yapo *et al.*, 1996).

The partitioning of the available field data between calibration and validation is not well reported in the literature. For example, Jewell *et al.* (1978) stated: "the verification data set should be as large as the calibration data set. However, if limited data are available, it is the usual practice to use the larger portion of the data for calibration and the smaller portion of the data for verification". We believe that it is of a particularly great importance to quantify terms such as "limited", "larger" and "smaller".

For that purpose, a method has been proposed to evaluate the effect of the number and the characteristics of sets of experimental data on both calibration and validation. In this paper, only SMC and EMC models will be presented.

Calibration of SMC models

For SMC models, there is no true calibration or validation. For each catchment, a set S with N measured rainfall events is available and assumed hereafter to be the whole available population of events. One proceeds to a random sampling from the population without replacement of subsets E_s, each one representing a hypothetical measurement campaign of n events, with $n \leq N$. Generating a large number of subsets (typically 1000 subsets) with the same number n of events allows to account for the variability of available data. Different sizes n of subsets are considered to take into account the effect of the number of events used for calibration. This leads to a statistical distribution of SMC values for each size of subsets. The generated distribution gives an indication on the error in the SMC value due to the number of measured events.

Calibration of EMC models

For EMC models, a similar sampling procedure is applied to build calibration subsets. The calibration of regression equations is done for each subset by linearization with a log10 transformation and ordinary least squares method. This choice has been made after tests with other calibration methods like direct search of an optimum.

Validation of EMC models

For each subset of n events used for calibration, all remaining $(N - n)$ events from the initial complete set E are used for validation. Consequently, the more events are used for calibration, the fewer events are available for validation.

Experimental catchments and data sets

To study the effect of the number and characteristics of measured events on calibration and validation of SQMs, large experimental data sets are required. The analysis is carried out for three pollutants: total suspended solids (SS), biological oxygen demand (BOD) and chemical oxygen demand (COD). Data from the French database QASTOR (Saget and Chebbo, 1996) and from "Le Marais" catchment in Paris, France (Chebbo *et al.*, 1999) are used. In the QASTOR data base, only 12 catchments with more than 9 measured EMCs have been selected. The catchments and the corresponding number of available data for each type of SQM are presented in Table 2 with event mean concentration values for SMC and EMC approaches and pollutographs for the BWMs. For example, there are $N = 41$ measured COD EMC values for the Aix-nord catchment.

Table 2 Selected urban catchments and number of available data for each modelling approach.

	SMC			EMC			BWM		SMC			EMC			BWM
Catchments	BOD	COD	SS	BOD	COD	SS	SS	**Catchments**	BOD	COD	SS	BOD	COD	SS	SS
Separated Sewer systems								**Combined sewer systems**							
Aix nord	38	48	48	33	41	41	12	Briche d11	16	17	17	12	13	13	17
Aix zup	43	52	52	41	47	47	20	Briche dd11	16	16	16	9	9	9	16
Ulis sud	31	31	31	29	29	29	10	Briche enghien	17	17	17	10	10	10	17
Velizy	27	27	27	22	22	22	19	Briche PHI	18	18	18	16	16	16	18
Maurepas	121	121	121	59	59	59	24	Briche PLB	18	18	18	16	16	16	18
Polluted separated Sewer system								Mantes	25	25	25	23	23	23	10
Ulis nord	75	75	75	57	57	57	14	Marais	-	64	67	-	64	67	42

Results and discussion

Evaluation of SMCs and EMCs for BOD, COD and SS has been carried out for all 13 catchments. As all results can not be presented here due to space limitations, only results for COD and SS for the "Le Marais" catchment are presented in the following paragraphs.

SMC models

The results for COD and SS are presented respectively in Figure 1 and Figure 2. The evolution of the SMC distribution for both pollutants varies with the number of events used for its evaluation. For example, when 1000 different subsets of $n = 10$ events are used, the mean value of COD SMC is approx. 315 mg/L, but COD SMC values have a 95 % probability to be between 237 and 410 mg/L. The limits of the 95 % confidence interval are not symmetric around the mean value, which reflects the fact that SMC distributions are log-normally distributed. Additionally, with $n = N = 64$ events, the 95 % confidence interval width is zero as the whole population has been used for calibration. We have neglected here the measurement uncertainties and the fact that the available population of measured events does not necessarily reflect the complete natural variability of the phenomena, as our objective is model calibration and not estimation of measurement uncertainties as e.g. given in Bertrand-Krajewski *et al.* (2002).

With 1000 subsets of $n = 40$ events, the mean value of COD SMC remains approximately equal to 310 mg/L, but the COD SMC values now range only from 281 to 340 mg/L with a 95 % probability. It appears clearly that increasing n leads to a better evaluation of the true (but unknown) SMC value. For both SS and COD, more than $n = 15$ events are needed to reach a stable mean value of the SMC distributions. However, by comparing Figure 1 and Figure 2, it appears that distributions are narrower for SS than for COD. In other words, for this catchment, COD values present a larger variability from event to event than SS values.

Figure 1 SMC and uncertainties related to the number of sampled events on "Le Marais" catchment for COD.

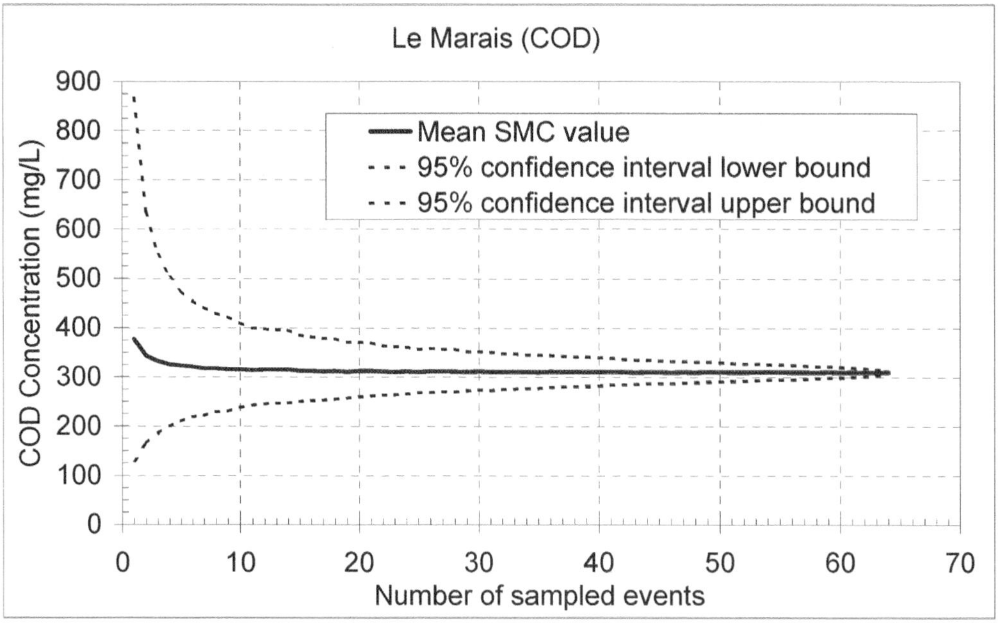

Additionally, it is important to note that practitioners usually do not have such large data sets. According to the survey described by Gromaire *et al.* (2002), only 87 % of the projects are carried out with calibrated models, and when models are calibrated, more than half of models are calibrated with less than 5 rainfall events. Considering $n < 5$ in Figures 1 and 2 reveals that SMC is evaluated with very large uncertainties and large bias.

For example, in Figure 1, for $n = 5$, the mean COD SMC value is approx. 323 mg/L with a 95 % probability to be between 210 and 475 mg/L depending of the subset used. This may have particularly dramatic consequences for operational purposes. Complementary analysis of all results for all catchments shows that the size N of the full data set is also critical, and that results depend strongly on it. However, a further analysis is needed to assess more accurately this influence and to evaluate if there are some general trends which would not depend on catchment characteristics or which would be specific for combined or separate catchments. More details can be found in Mourad *et al.* (2004).

Figure 2 SMC and uncertainties related to the number of sampled events on "Le Marais" catchment for SS.

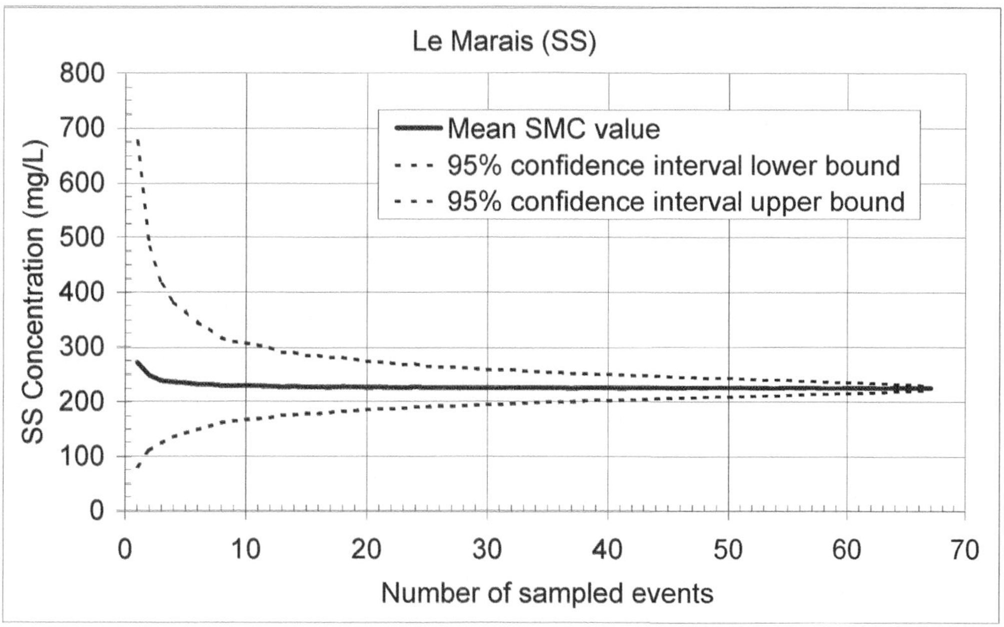

EMC models

EMC regression models have been calibrated for all calibration subsets. The Root Mean Squared Error (RMSE) is used as an indicator of goodness for both calibration and validation. Figure 3a represents the probability of having a calibration RMSE value as given on the bottom axis for the Servat model applied for suspended solids, for various sizes of calibration subsets. For example, by using $n = 40$ rainfall events for calibration, there is a probability of approx. 47 % to reach a calibration RMSE = 105. The value of RMSE = 105 corresponds to the most likely calibration RMSE value when all N rainfall events are used for calibration. For $n = 4$, one observes a probability equal to 1 of having a calibration RMSE equal to zero. In other words, the model fits perfectly the experimental data. This is not surprising, as it is easy to perfectly fit 4 rainfall events with a model having 4 calibration parameters. One also observes that about 25 events are needed to achieve a centred distribution around calibration RMSE = 105. In other words, for this catchment and this model, it is recommended to use a minimum of approx. $n = 25$ events for calibration in order to get a calibration quality which has a significant probability to be close to the best achievable calibration obtained when all N rainfall events are used. With $n \geq 45$ events, this probability becomes higher than 50 %.

Figure 3b represents the probability of having a validation RMSE value as given on the bottom axis for the Servat model applied for suspended solids, for various sizes of calibration subsets. For $n = 4$, the validation RMSE values are very high and located out of the figure for scale purposes. Moreover it seems that a centred distribution around 100 is achieved for calibration using around $n = 15$ events. The probability of having satisfactory values of validation RMSE is highest for n between 20 and 35 events. The most probable values of RMSE can be observed around 105 for calibration and around 100 for validation. The difference is quite small and means that validation is slightly better than calibration, which is unusual. Figure 4 presents the results for COD.

Figure 3 RMSE distribution related to the number of sampled events for SS

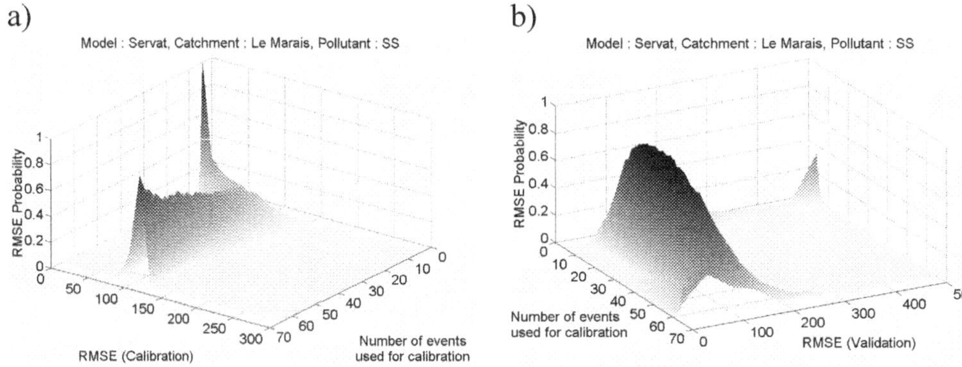

Figure 4 RMSE distribution related to the number of sampled events for COD

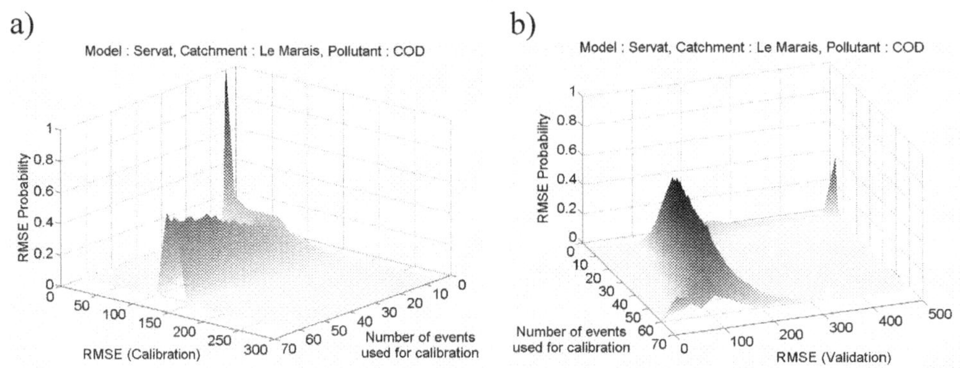

Similar conclusions can be drawn as for SS in Figure 3. But unlike Figure 3, the most probable values of RMSE are around 125 for calibration and around 150 for validation: calibration is now better than validation, which is the most frequent case. The distribution of

RMSE values for the SS model calibration is sharper than for COD. Consequently, the calibration of the Servat model for SS is less sensitive to data characteristics than for COD. Nevertheless, for both COD and SS, the size *n* of calibration data sets is critical. Validation figures show that, for COD, the number of events used for the validation influenced the RMSE distribution. Most probable values and a rather stable distribution are observed for *n* between 15 and 20. For SS, the size *n* of validation data sets does not influence much the RMSE distribution. Most probable values and a rather stable distribution are observed for *n* between 20 and 40 events, or (*N-n*) between 27 and 47 events.

Conclusions

A method to assess the uncertainty in SQMs calibration and validation due to the variability in calibration data has been presented and applied to 13 urban catchments having different physical and land-use characteristics. Preliminary results for SMC and EMC models applied to "Le Marais" catchment for SS and COD are presented and show that the size *n* of calibration data sets is very critical. With $n < 15$ to 20 events, the results might be extremely biased and the consequences on management and design studies can be dramatic. The results demonstrate the significance of the model validation, which is frequently neglected by practitioners. A good calibration doesn't necessarily lead to a good prediction ability of the model: a validation with independent data sets must be performed. The size *n'* of the validation data sets appears as less critical than for calibration, provided that a minimum size is used. For *n'* between 20 and 40 events, the validation quality remained rather constant for SS and showed little variations for COD. These conclusions strongly depend on the available data sets. Nevertheless, the observed trends are similar for all 13 catchments which have been studied. A further detailed analysis of all results will be carried out in order to evaluate i) the content of information included in calibration and validation data subsets, and ii) the potential similarities that could be linked to the type of catchments and sewers (combined and separate). It is important to mention that the results are relative to the number of available data sets and don't take into account measurement uncertainties.

Acknowledgements

The authors gratefully acknowledge the financial support of the RGCU "Réseau Génie Civil et Urbain".

References

Ashley R.M., Hvitved-Jacobsen T., Bertrand-Krajewski J.-L. (1999). Quo vadis sewer processes modelling ? *Water Science and Technology*, **39**(9), 9-22.

Bertrand-Krajewski J.-L., Barraud S., Bardin J.-P. (2002). Uncertainties, performance indicators and decision aid applied to stormwater facilities. Urban Water, 4(2), 163-179. ISSN 1462-0758.

Bouteligier R., Vaes G., Berlamont J. (2002). In sewer sediment transport models. Proceedings of the 3rd International Conference on Sewer Processes and Networks, Paris, France, 15-17 April 2002, 47 – 54.

Calabrò P.S., Maglionico M. (2002). Comparison between a detailed model and two simplified models in urban stormwater quality simulation. *Proceedings of the 9th International Conference on Urban Drainage (ICUD)*, Portland, USA, 8-13 September 2002, 14 p.

Chebbo G., Gromaire M.-C., Garnaud S., Gonzalez A. (1999). The Experimental Urban Catchment "Le Marais" in Paris. *Proceedings of the 8th International Conference on Urban Storm Drainage (ICUSD)*, Sydney, Australia, 30 August - 3 September 1999, 1520-1527.

DHI (2000). MOUSE TRAP technical reference. Hørsholm, Denmark.

Driver N.E., Tasker G.D. (1990). *Techniques for estimation of storm-runoff loads, volumes, and selected constituent concentrations in urban watersheds in the United States.* Washington, DC. (USA), U.S. Geological Survey, Water-Supply Paper 2363, 44 p.

Gromaire M.-C., Cabane P., Bertrand-Krajewski J.-L., Chebbo G. (2002). Operational use of urban drainage pollutant flux models - results from a French survey. *Proceedings of the international conference on Sewer Operation and Maintenance SOM 2002*, Bradford, UK, 26-28 November 2002, 8 p.

Hoos A.B., Sisolak J.K. (1993). *Procedures for adjusting regional regression models of urban-runoff quality using local data.* Nashville, Tennessee (USA) : USGS, USGS open-file report 93-39, 39 p.

INSA/SOGREAH (1999). *CANOE User Manual.* Villeurbanne (France) : ALISON, INSA de LYON, 1999.

Jewell T.K., Nunno T.J., Adrian D.D. (1978). Methodology for calibrating stormwater models. *Journal of the Environmental Engineering Division*, **104**(3), 485-501.

Kanso A., Gromaire M.-C., Gaume E., Tassin B., Chebbo G. (2002). Bayesian approach for the calibration of models : application to an urban stormwater pollution model. *Water Science and Technology*, **47**(4), 77–84.

Mourad M., Bertrand-Krajewski J.-L., Chebbo G. (2004). Uncertainty in averaging mean event concentrations of storm water pollution according to the number of measured events. *Proceedings of NOVATECH 2004*, Lyon, France, 6-10 June 2004, vol. 1, 705-712.

Saget A. (1994). *Base de données sur la qualité des rejets urbains de temps de pluie :distribution de la pollution rejetée, dimensions des ouvrages d'interception.* PhD Thesis, Ecole Nationale des Ponts et Chaussées, Paris (France), 333 p.

Saget A., Chebbo G. (1996). Qastor: The French Database about the Quality of Urban Wet Weather Discharges. *Proceedings of the 7th International Conference on Urban Storm Drainage (ICUSD)*, Hannover, Germany, 9-13 September 1996, vol. 3, 1707-1713.

Schlütter F. (1999). *Numerical modelling of sediment transport in combined sewer systems.* PhD thesis, Aalborg University, Aalborg, Denmark, 172 p.

Schueler T.R. (1987). *Controlling Urban Runoff : A Practical Manual for Planning and Designing Urban BMPs.* Washington, DC. (USA) : Metropolitan Washington Council of Governments, Publication No. 87703, 202 p.

Servat E. (1984). *Contribution à l'étude de des matières en suspension du ruissellement pluvial à l'échelle d'un petit bassin versant.* Ph.D. thesis, Université des Sciences et techniquesdu Languedoc, France, 182 p.

Yapo P., Gupta H.V., Sorooshian S. (1996). Automatic Calibration of Conceptual Rainfall-Runoff Models: Sensitivity to Calibration Data. *Journal of Hydrology*, 181, 23-48.

Zug M., Bellefleur D., Phan L., Scrivener O. (1997). Sediment transport model in sewer networks, a new utilisation of Velikanov model. *Proceedings of the 2nd International Conference on the sewer as a physical, chemical and biological reactor*, Aalborg, Denmark, 25-28 May 1997, 8 p.

Zug M. (1998). *Modélisation conceptuelle de la pollution par temps de pluie en réseau d'assainissement.* Thèse de doctorat de l'Université Louis Pasteur, Strasbourg, 241 p.

Prediction of Sediment Deposition in Raja River Concrete Drainage System: A Case Study

Kassim, M., Ab. Ghani, A., Abdullah, R., Zakaria, N.A.

River Engineering and Urban Drainage Research Centre (REDAC),
Engineering Campus, Universiti Sains Malaysia, 14300 Nibong Tebal, Penang, MALAYSIA

Abstract The existing conventional drainage system in cities throughout Malaysia is built mainly to cater for the increase in surface runoff due to rapid development that occurs in the city borders. The drainage system is normally made up of open rigid concrete drains susceptible to maintenance problems such as sediment deposition, litter, and utility pipes crossing and blocking the waterways. The paper highlights the result of the survey carried out to determine the sediment size characteristics in urban drains throughout cities in Malaysia. New result from sediment deposition profile measured at Raja River drainage system, Alor Setar is also described. Assessment of the existing incipient motion criteria was also made to identify the equation that is able to predict the sediment deposition trend in the River Raja drainage system.

Keywords Sediment deposition; self-cleansing; sediment size; storm drain

Introduction

Rapid development in major towns in Malaysia in the last decade results in the construction of new drainage systems mainly, open monsoon or storm drains, to cater for the increase in surface runoff. While the major function of these drains is to ensure rapid conveyance of surface runoff to avoid flash flood, sediment deposition of the drains seems to be ignored. As a result, significant deposition occurs even in new storm drains resulting in higher maintenance costs to regain the designed hydraulic capacity of the drains.

The Raja drainage catchment is part of the Alor Setar Town Drainage scheme, Phase I. The scheme comprises three catchment basins (Raja, Langgar & Putera) with a total catchment area of 300 ha and was proposed as part of a flood alleviation programme for the region. The Raja basin is the largest in the area (233 ha) and is also the most prone to problems of flooding. The area receives an annual rainfall of 2500 mm. Flows from the Raja basin are drained by gravity to a pump station before being discharged via pumping mains to the Sungai Kedah waterway. The main lengths of the drainage system take the form of trapezoidal/rectangular, open channel sections. The sections are lined in concrete, vary in width from 4 to 16m and have been designed to a minimum velocity criterion of 0.9 m/s for the purpose of sediment self-cleansing.

A study was carried out in 1997 to identify the sediment size characteristics along the River Raja channel (Ab. Ghani et al., 2000). Kassim (2003) conducted new data collection of sediment sizes in urban drains at several main cities in Malaysia. Further data collection was

 Sewer Networks and Processes within Urban Water Systems edited by J.-L. Bertrand-Krajewski, M. Almeida, J. Matos & S. Abdul-Talib. ISBN: 1 84339 506 1.

also made along Raja River drainage system to identify the trend of sediment depositions. This paper will highlight the result of this new data collection programme.

Previous Studies

The self-cleansing approach in terms of either minimum velocity (V_c) or shear stress (τ_c) is usually used to minimize sediment problems in urban drainage systems. The application of this approach is meant to avoid any deposition at any time or at least over a long period of time no deposits will build up (i.e. high flows flush the deposits). Ab. Ghani (1997) highlighted problems encountered in Europe in applying a constant self-cleansing approach normally in terms of either velocity or shear stress. Recent studies (Ab. Ghani et al. 1999, Nalluri & Ab. Ghani, 1996, Nalluri et al. 1994) suggest new methods to improve the constant velocity approach method by taking into account several factors affecting sediment transport, namely sediment characteristics (size, sediment concentration) and drain characteristics (size, slope and roughness).

Incipient motion criteria are also being used to design self-cleansing sewers. Table 1.1 shows a list of available incipient motion criteria used to predict sediment deposition along the Raja River channel (Novak & Nalluri 1975, Ojo 1978, Novak & Nalluri 1984, El-Zaemey 1991, Ab.Ghani et al. 1999).

Table 1.1 Summary of Incipient Motion Criteria

Researchers	Incipient Motion Criteria	Equation No.
Novak & Nalluri (1975)	$Vc = 0.17(S_s-1)^{1/2} d_{50}^{0.24}$	1
Novak & Nalluri (1984)	$Vc / [gd_{50}(Ss - 1)]^{1/2} = 0.5 [d_{50}/R]^{-0.4}$	2
Novak & Nalluri (1975)	$Vc / (gd_{50})^{1/2} = 0.61(Ss - 1)^{1/2} (d_{50}/R)^{-0.27}$	3
El-Zaemey (1991)	$Vc / [gd_{50}(Ss - 1)]^{1/2} = 0.75 [d_{50}/R]^{-0.34}$	4
Ojo (1978)	$Vc = 0.184(Ss - 1)^{1/2} d^{0.24}$	5
Ojo (1978)	$Vc = 0.65(gd)^{1/2}(Ss - 1)^{1.2}(d/R)^{-0.28}$	6
El-Zaemey (1991)	$Vc / [gd_{50}(Ss - 1)]^{1/2} = 0.80 [d_{50}/R]^{-0325}(y_o/B)^{0.04}$	7
Ab.Ghani et al. (1999)	$Vc / [(gd_{50})(Ss - 1)]^{1/2} = 1.07(d_{50}/R)^{-0.23}$	8
Novak & Nalluri (1975)	$\tau c = 0.128(Ss-1)d_{50}^{0.4}$	9
Ojo (1978)	$\tau c / \rho gd_{50}(Ss-1) = 0.062(U^*/\nu d_{50})^{-0.54}$	10
Ab.Ghani et al. (1999)	$\tau c / [\rho gd_{50}(Ss-1)]^{1/2} = 0.17 D_{gr}^{-0.57}$	11

Sediment Size Characteristics

A survey of sediment size characteristics was made at several concrete drains throughout main cities in Malaysia namely Northern Region (Kangar, Alor Setar, Butterworth, Ipoh), Central Region (Kuala Lumpur, Shah Alam, Seremban, Malacca), Southern Region (Johore Bahru), and East Cost Region (Kuantan, Kuala Terengganu, Kota Bharu).

The sediment samplings were made by grab. Dry sieving was made to obtain the sediment distribution curves for these cities as shown in Figure 1.1. It shows that the average sediments size in the urban drainage system for all cities is 0.7mm.

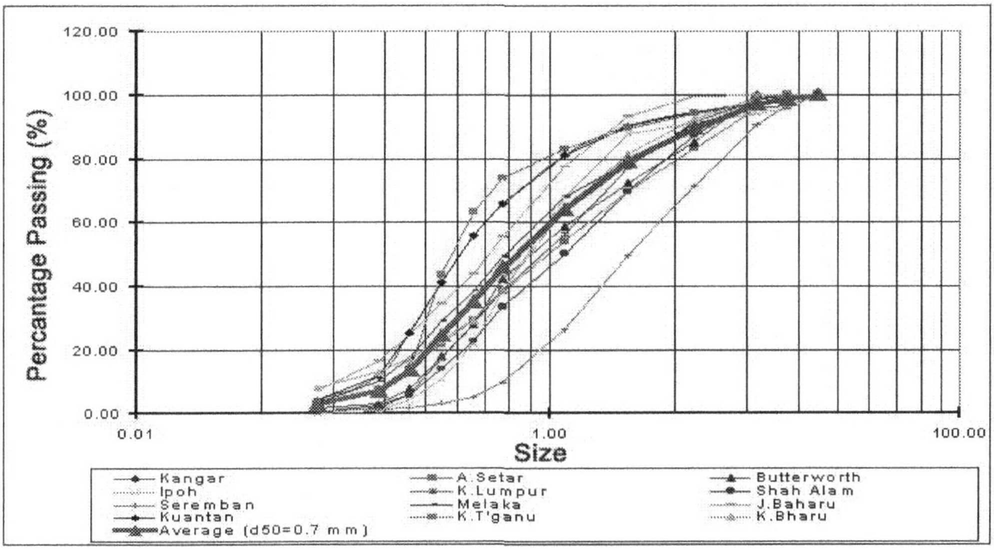

Figure 1.1 Sediment Distribution Curves for Main Cities in Malaysia

Sediment Deposition Trend Along Raja River Drainage System

A new sediment data collection programme (Kassim, 2003) was carried out in 2000 whereby measurement of sediment deposition were made at ten stations along Alor Derga and Alor Siam forming the upstream branches and the main reach of the Raja River (Figure 1.2).

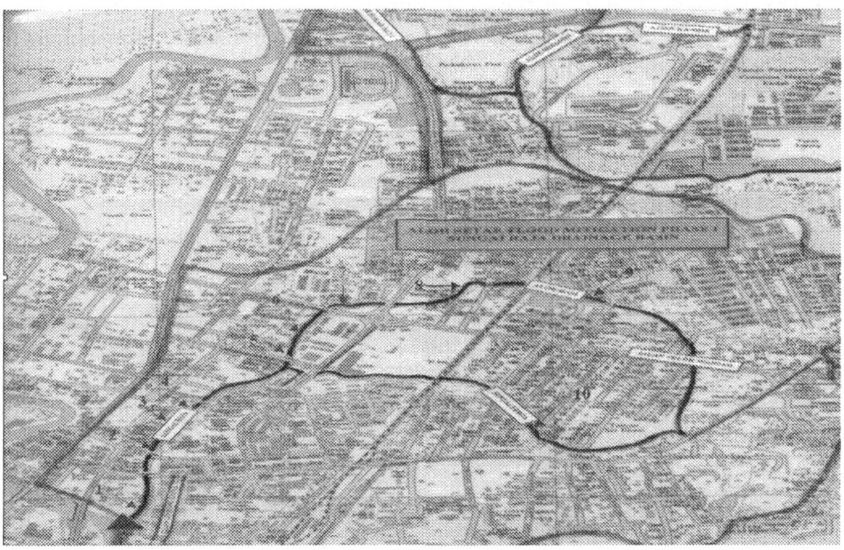

Figure 1.2 Raja River Drainage System

Sediment Profile Measurement

For each station, the thickness of sediment deposits along the channel was measured over a 20-metre distance at an interval of 1m spacing. For each interval, three measurements of sediment thickness were made (Figure 1.3). Samples of sediment were also collected by grab during the measurement of sediment profiles.

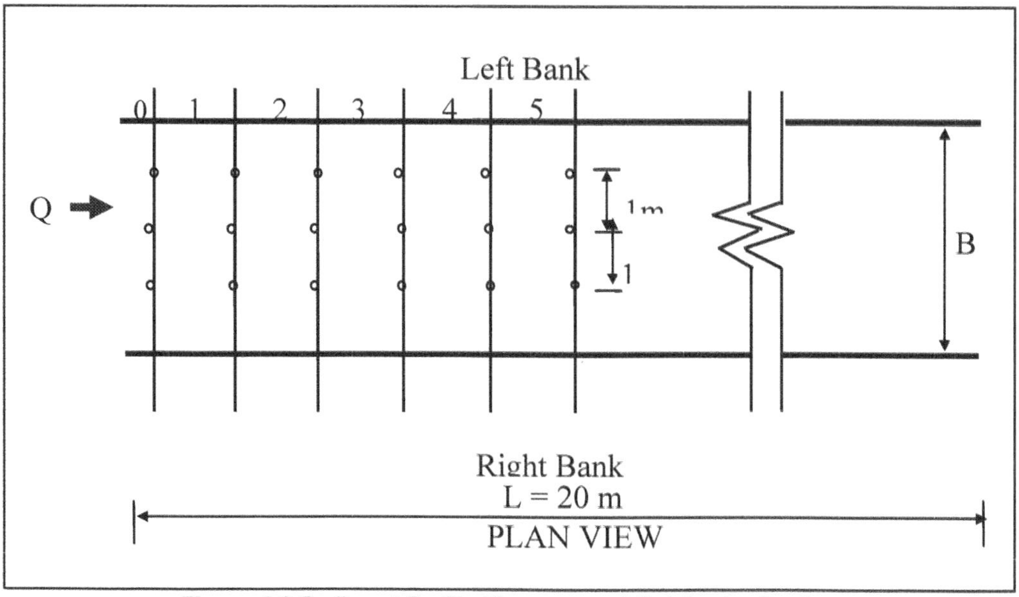

Figure 1.3 Sediment Profile Measurement Points at a Section

The sediment distribution curves for all stations are shown in Figure 1.4. It shows that the average sediment size in the Raja River drainage system is 0.8mm.

An example of sediment profile for Station 1 measured on 10th June 2000 is given in Figure 1.5. For the main channel, it shows that the deposition thickness varies between 0.2m and 0.4m. The average sediment profile with a slope of 0.0008 as compared to the main channel slope of 0.0004 is given in Figure 1.6.

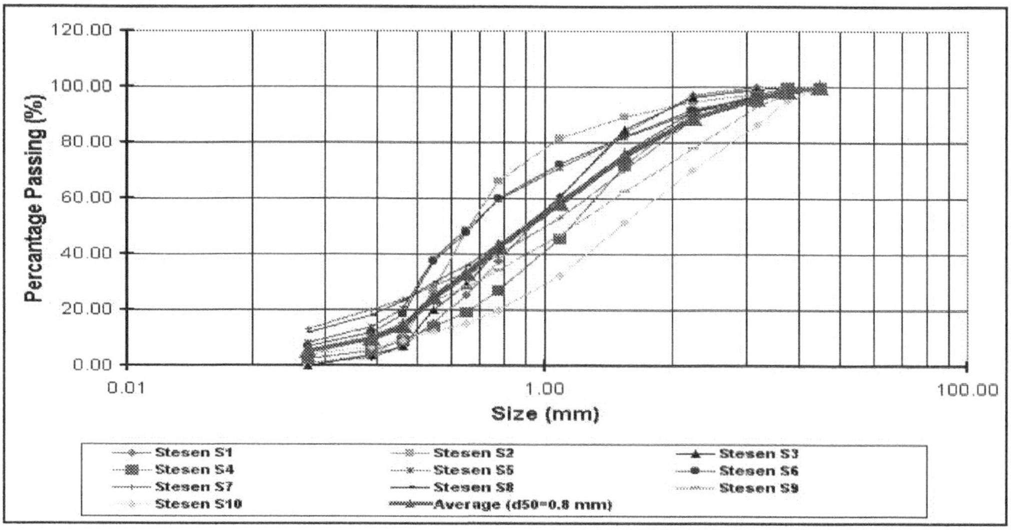

Figure 1.4 Sediment Distribution Curves for All Stations, Raja River

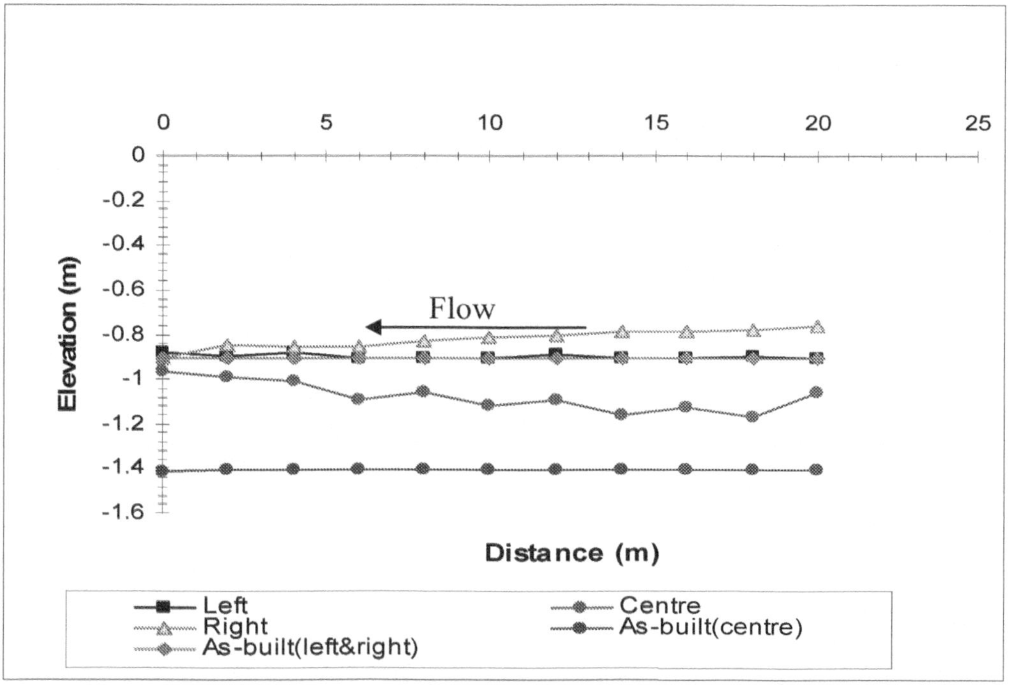

Figure 1.5 Sediment Deposition Profile for Station 1 (10th June 2000)

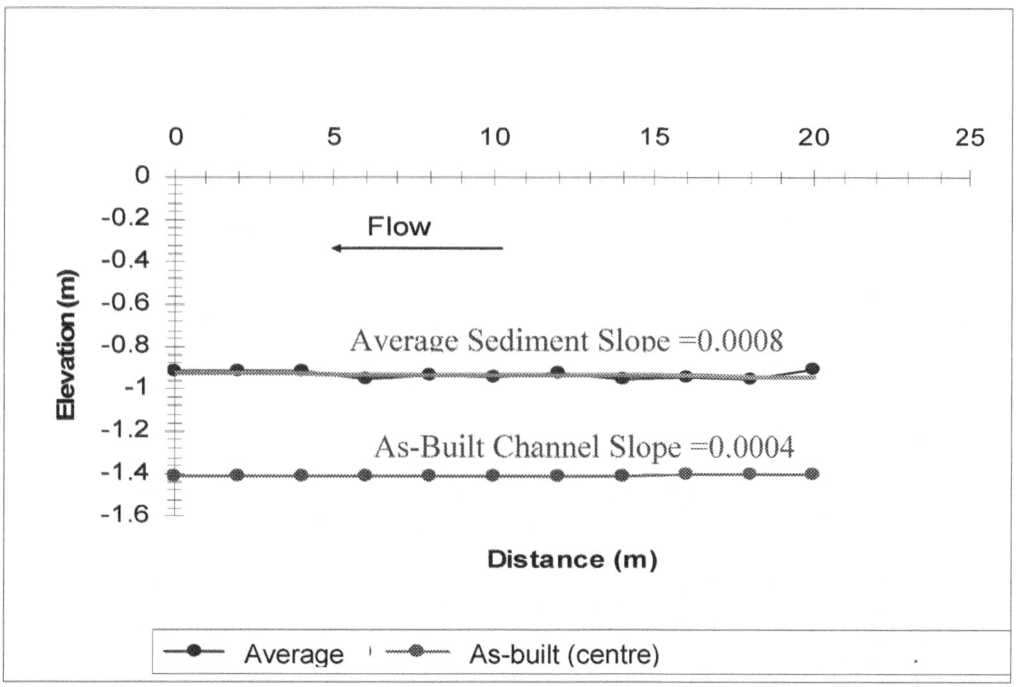

Figure 1.6 Average Sediment Deposition Profile for Station 1 (10th June 2000)

Assessment of Existing Incipient Motion Criteria

The comparisons of the sediment deposition for Station 1 by these different criteria (Table 1.1) are shown in Table 1.2 and Figure 1.7. Table 1.2 also gives the measured and as-built slopes for comparison purpose. The critical velocity criterion given by Ab. Ghani et al. (1999) seems to give the best prediction of the as-built slope while the shear stress criterion of Ab. Ghani et al. (1999) is able to give the nearest value of the measured slope of the sediment deposits.

Table 1.2 Computed Slopes vs. Measured and As-Built Slopes

Researchers	Equation No.	Computed Slope
Novak & Nalluri (1975)	1	0.000047
Novak & Nalluri (1984)	2	0.00095
Novak & Nalluri (1975)	3	0.00023
El-Zaemey (1991)	4	0.00093
Ojo (1978)	5	0.000002
Ojo (1978)	6	0.00061
El-Zaemey (1991)	7	0.00070
Ab.Ghani et al. (1999)	8	0.00041
Novak & Nalluri (1975)	9	0.000023
Ojo (1978)	10	0.000008
Ab.Ghani et al. (1999)	11	0.0013
As-Built Slope		0.0004
Measured Slope		0.0020

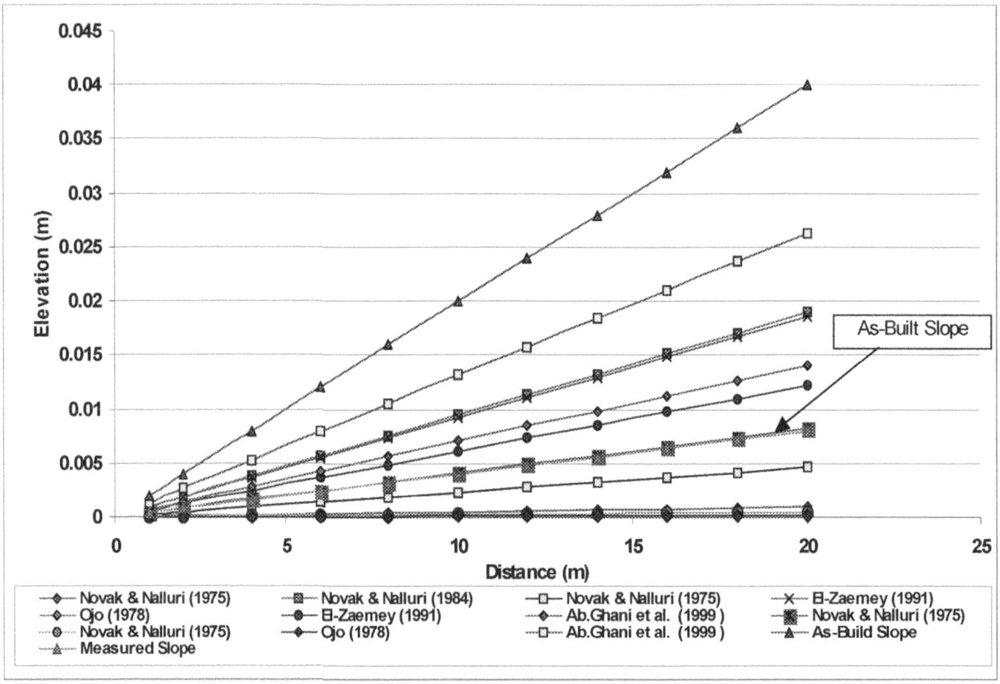

Figure 1.7 Appraisal of Existing Incipient Motion Equations for Raja River, Alor Setar (Station 1, 10th June 2000)

Ab. Ghani et al. (1999) developed their equations for incipient motion under steady flow conditions in a rectangular channel (Salem, 1998). Four different bed thicknesses up to 24 mm of sediment deposition were studied for a range of sediment size between 0.5 mm and 4.78 mm. The good agreement obtained for the equations by Ab. Ghani et al (1999) appears to be the results of including different depths of sediment deposition in their study.

The data collection was made before and after a monsoon season in 2000. The results show that the supply of sediment to the drainage system would determine the deposition trend along Raja River drainage system.

Conclusions

A survey of sediment size characteristics was made at several concrete drains throughout main cities in Malaysia. The results show that the average sediment size is 0.7mm. A detailed study on sediment deposition trend was carried out in 2000 whereby measurements of sediment deposition were made at ten stations along Raja River drainage system in Alor Setar. An assessment of several incipient motion criteria shows that taking into account the sediment deposition thickness in the equation developed e.g.. Ab. Ghani et al. (1999) will give good agreement with the measured data.

Reference

Ab. Ghani, A. (1997) Field data collection on sedimentation in combined sewer systems: overseas experience. *Bulletin, Institution of Engineers, Malaysia*. **11**, 40-47.

Ab. Ghani, A., Zakaria, N.Z., Kassim, M., and Ahmad Nasir, B. (2000) Sediment size characteristics of urban drains in Malaysian cities. *Urban Water*. **2**(4), 335-341.

Ab. Ghani, A., Salem, A.M., Abdullah, R., Yahaya, A. S., and Zakaria, N.A. (1999) Incipient motion of sediment particles over deposited loose beds in a rectangular channel. In *Proc. Of the 8th Int. Conf. on Urban Storm Drainage*, Sydney, 30 August– 3 September.

El-Zaemey, A.K.S. (1991) Sediment transport over deposited beds in sewers. Ph.D. Thesis, Dept Civil Eng., University of Newcastle upon Tyne, England.

Kassim, M. (2003). Sediment Deposition in a Rigid Monsoon Drain along Raja River Catchment, Alor Setar. M.Sc. Progress Report No.3, Dept Civil Eng., University Of Science, Penang, Malaysia.

Novak, P & Nalluri, C. (1975). Sediment transport in smooth fixed bed channels. *Hydraulics Division*. **101**(HY9). 1139-1154.

Novak, P & Nalluri, C. (1984). Incipient motion of sediment particles over fixed beds. *Hydraulic Research*. **22**(3), 181-197.

Ojo, S.I.A. (1978). Study of incipient motion and sediment transport over fixed beds, Ph.D. Thesis, Dept Civil Eng., University of Newcastle upon Tyne, England.

Salem, A. M. (1998). Incipient motion of sediment particles over loose deposited beds in a rectangular channel. M.Sc. thesis, Dept Civil Eng.,Univ. of Science, Penang, Malaysia.

Experimental validation of the DORA model for bed load transport in storm sewers

C. Aricò*, R. Gaudio, A. Miglio*****

* Dipartimento di Ingegneria Idraulica ed Applicazioni Ambientali, Università di Palermo, Viale delle Scienze, 90128 Palermo, Italy, Tel.: +39-091-6657.73, Fax: +39-091-6657.49, E-mail: arico@idra.unipa.it

** Istituto di Ricerca per la Protezione Idrogeologica (IRPI) – Sezione di Cosenza, Consiglio Nazionale delle Ricerche (CNR), Via Cavour, 4-6, 87036 Rende (CS), Italy, Tel.: +39-0984-835251, Fax: +39-0984-835319, E-mail: gaudio@irpi.cnr.it

*** Dipartimento di Difesa del Suolo, Università della Calabria, 87036 Arcavacata di Rende (CS), Italy, Tel.: +39-0984-496554, Fax: +39-0984-494050, E-mail: a.miglio@dds.unical.it

Abstract. The DORA (Double ORder Approximation) model has been extended to the solution of 1D shallow water equations in unsteady-state flow and mobile bed condition. The DORA methodology can be classified among the fractional time step methods. Results of the numerical code have been compared to experimental data obtained in unsteady non equilibrium conditions using a semi-ovoid perspex pipe in the laboratory flume of the Laboratory of Hydraulic Models in the Department of Soil Defence, University of Calabria, Italy. Cases of degradation have been considered. Results of the model seem to fit profiles of sediment and water depth quite well using bed load versus velocity relationships already proposed in literature for equilibrium conditions.

Keywords Bed load; degradation; numerical models; sediment transport.

INTRODUCTION

Several unsteady mobile boundary numerical models have been developed during the last few decades to simulate bed level changes and suspend solid transport in natural and artificial channels (Correia et al., 1992; Kassem and Chaudhry, 1998; Cao et al., 2002).

Recently, the numerical model DORA (Tucciarelli and Termini, 2000; Noto and Tucciarelli, 2001; Aricò and Tucciarelli, 2002, 2003) has been introduced to predict sediment transport in unsteady flow conditions. In this paper characteristics of the DORA model are recalled, as well as the capacity of the model to reproduce experimental data measured in the laboratory flume of the Laboratory of Hydraulic Models in the Department of Soil Defence, University of Calabria, Italy.

Equations and parameters of the DORA Model

The DORA model can be listed among the fractional time step methods (Tucciarelli and Termini, 2000; Noto and Tucciarelli, 2001). The solution of the original system, from time level k to time level k+1, is obtained by sequentially solving one prediction and one correction system.

It is assumed that solid material moves as bed load and dissolved substances move as

 Sewer Networks and Processes within Urban Water Systems edited by J.-L. Bertrand-Krajewski, M. Almeida, J. Matos & S. Abdul-Talib. ISBN: 1 84339 506 1.

suspended load, without affecting the water density. The thickness of the bed layer, δ in Fig. 1, is assumed negligible with respect to the water section depth.

The proposed DORA model solves a system of Partial Differential Equations (PDEs) given by continuity and momentum equations of the liquid phase as well as the continuity equation of the suspended and bed load. The unknowns of the problem are water depth h, mean flow velocity U, depth averaged suspended solid concentration C and bed level z_b. The diffusive hypothesis is considered in the momentum flow equation. This is equivalent to assuming an infinite gravity force. Differences between results obtained in the "fixed bed" case using the complete and the diffusive model is negligible in most of practical application where the inertial forces do not have a major impact, like dam-break waves close to the origin or short-length waves in lagoons and estuaries. By merging the continuity and the momentum equation for the water phase and applying the Chezy formula for flow resistance, the governing PDEs system becomes:

$$\frac{\partial \sigma}{\partial t} - \frac{\partial}{\partial x}\left(\sigma \chi S \sqrt{R/|S|}\right) = 0 \qquad (1),$$

$$\frac{\partial (C\sigma)}{\partial t} + \frac{\partial (QC)}{\partial x} - K\frac{\partial^2 C}{\partial x^2} = 0 \qquad (2),$$

$$(1-p)\frac{\partial z_b}{\partial t} + \frac{\partial q_s}{\partial x} = 0 \qquad (3),$$

where t is time, x the coordinate in the streamwise direction, σ the wetted area, R the hydraulic radius, Q the flow rate, S the flow resistance per unit water weight, p the bed porosity, χ the Chezy coefficient, q_s the total solid load in absolute volume per unit width and K the dispersion coefficient, estimated as a function of mean velocity, U, water depth, h, and shear stress (Fisher et al. 1979).

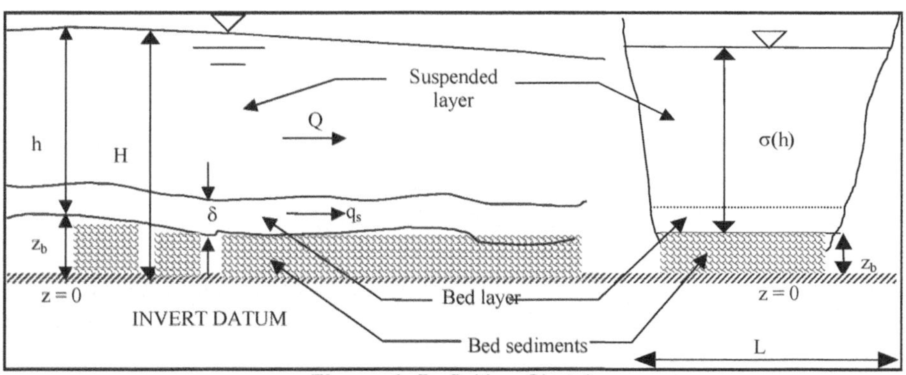

Figure 1. Definition Sketch

Several formulations have been proposed in literature for bed load transport; this is considered function of hydrodynamic variables, essentially the water depth and mean flow velocity, bed shear-stress and sediment characteristics, according to coefficients estimated during equilibrium conditions. The formula proposed by Ackers and White (1973), afterwards modified by Ackers (1984), has been used for bed load transport simulation:

$$q_s = C_V \sigma U \tag{4}$$

where C_V is the volumetric sediment concentration:

$$C_V = G_{g,r} \Big/ \left(\frac{R}{d_{50}}\right)\left(\frac{\sigma}{W_e R}\right)^{1-n}\left(\frac{u^*}{U}\right)^n \tag{5}$$

d_{50} the mean size of the granulometric distribution, W_e the bed width, $u^* = \sqrt{gRS}$ the shear velocity and $G_{g,r}$ the dimensionless sediment transport parameter:

$$G_{g,r} = \overline{H}\left(\frac{F_{g,r} - A_{g,r}}{A_{g,r}}\right)^m \tag{6}$$

depending on the mobility parameter $F_{g,r}$ given by:

$$F_{g,r} = \frac{u^{*n}}{\sqrt{g(s-1)d_{50}}} U^{1-n} \Big/ \left(\sqrt{32}\log_{10}\frac{12R}{d_{50}}\right)^{1-n} \tag{7}$$

$A_{g,r}$ represents the mobility parameter at the movement threshold. The coefficients reported in the above equations are related to the dimensionless grain size $D_{g,r} = d_{50}\left[g(s-1)/\nu^2\right]^{1/3}$ according to:

$$A_{g,r} = 0.14 + 0.23\Big/\sqrt{D_{g,r}}, \quad m = 1.34 + 9.66/D_{g,r}, \quad n = 1.00 - 0.56\log_{10} D_{gr},$$

$$\log_{10}\overline{H} = 2.86\log_{10} D_{g,r} - \left(\log_{10} D_{g,r}\right)^2 - 3.53 \tag{8}$$

where s is the relative density of sediment with respect to the water and ν is the water kinematic viscosity. The formula has been tested for the following range of variation of dimensionless grain size: $1<D_{g,r}\le 60$. For coarser sediments, i.e. $D_{g,r}>60$, good results can still be obtained by setting $n=0$, $A_{g,r}=0.17$, $m=1.5$ and $\overline{H} = 0.025$.

More details on the solution of the prediction and correction steps in Noto and Tucciarelli, (2001), Aricò and Tucciarelli (2002, 2003). The proposed DORA scheme has been proven by means of the Fourier analysis to be unconditionally stable in the linear case, for example in the solution of the groundwater convective transport problem (Tucciarelli and Fedele, 2000). Unconditional stability has been empirically observed in several non-linear 1D and 2D applications (Tucciarelli and Termini, 2000; Noto and Tucciarelli, 2001; Aricò and Tucciarelli, 2002).

Experimental installation

Experiments were carried out at the Laboratory of Hydraulic Models of the Department of

Soil Defence, University of Calabria, Italy. The facility was an 11.0m-long semi-ovoid perspex channel (250 mm × 375 mm ovoid cross-section), bolted to a tilting truss with longitudinal slope variable from 0.00 to 5.00 %.

The channel bottom was divided in three sections, as shown in Fig. 2: the 3.65m-long end sections presented a fixed bed, whereas the 3.7m-long middle section a mobile bed, with maximum sediment depth 0.047 m. The sediment used in tests was a uniform fine gravel, characterised by a median diameter d_{50}=3.0 mm and a specific weight γ_s=26.5 kN/m^3.

The end sections were realised with 130 mm-wide and 4.0 mm-thick PVC strips, set at the distance of 40 mm from the channel invert. In order to obtain a uniform roughness along the whole channel, a layer of the aforementioned sediment was glued on the PVC strips.

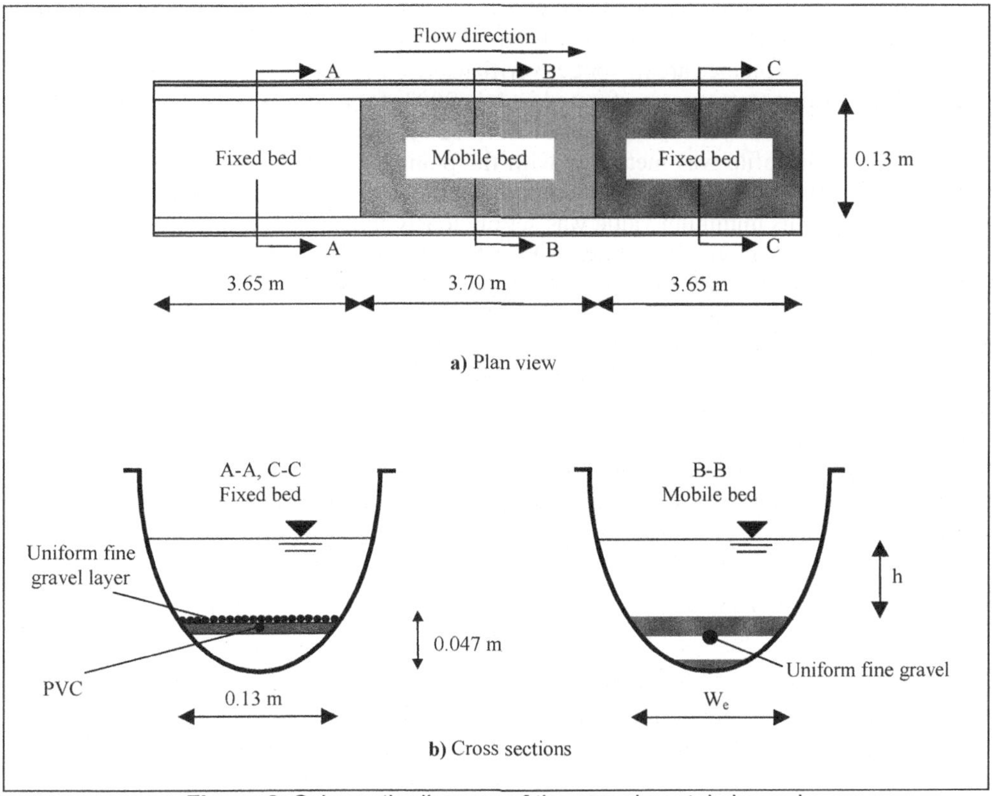

Figure 2. Schematic diagram of the experimental channel.

The initial longitudinal bed profile in the middle section was marked with some adhesive transparent strips. In this way it was possible to observe the variations of the bed elevations during the runs.

The channel was fed by a pump submerged into a sump through a 100 mm diameter pipe; the water flowed into a tank at the upstream end of the channel; a grid was inserted to minimise flow disturbances at the channel inlet. The discharge was measured with an ultrasonic flowmeter mounted on the feeding pipe.

A restitution channel, with a Thomson weir for calibration, leading to the sump for the re-circulation, receives the water flowing from the downstream end of the experimental channel.

Water depths were measured in several cross-sections with a digital point gauge with accuracy of ±0.01 mm.

Sediments were introduced into the channel through an Archimedes's screw fed by a 70 dm^3 hopper, followed by a chute. The screw is driven by a potentiometer, which permits setting a constant value of the screw rotation velocity; therefore, a uniform volumetric sediment discharge can be injected. The potentiometer was accurately calibrated versus solid discharge. The sediment transported by the flow was collected in a basket at the experimental channel outlet, properly dried and weighed.

Diagnostic tests

The research aim was the investigation of the degradation phenomenon in the mobile bed channel reach, for given liquid discharge and slope values. In fact, degradation occurs once the sediment feeding at the channel inlet is stopped and the equilibrium conditions between erosion and deposit volumes are no longer verified.

Three degradation tests were performed in July 2003, for three different discharges Q. The longitudinal slope S was constant and equal to 1.122 %.

The initial bed profile was measured with the point gauge at the beginning of each run. Then the liquid discharge was gradually pumped into the channel, until a minimum flow rate was established. This minimum value was set obviously higher than the value corresponding to incipient sediment motion condition. Once the motion threshold was exceeded and the solid particles were regularly transported by the flow, the sediment feeding was increased along with the liquid discharge. This avoided erosion along the middle channel and especially local scouring at the transition between fixed and mobile beds. Once the final Q value was reached, the solid discharge was increased or decreased to achieve equilibrium between erosion and deposition. The equilibrium condition was identified by comparing the solid discharge fed at the inlet, $q_{s,in}$, with the average solid discharge at the outlet, $q_{s,out}$, computed as the ratio between the weight of the dried material collected at the outlet in 1 minute and the time interval Δt=1 min. The equilibrium condition was therefore expressed as follows:

$$\left(\left|q_{s,in} - q_{s,out}\right|\right)/q_{s,in} < 0.05 \qquad (9)$$

At equilibrium, the transparent adhesive strips allowed the observation of the steady-state bed profile. In addition, the uniform flow condition was verified by measuring the hydraulic depths in the mobile bed channel section. In Table 1 the flow depth h_0, flow rate Q_0, mean flow velocity U_0, bed slope S_0, bed load q_{s0} and Darcy-Weisbach friction factor f_0 under the initial equilibrium state ($f_0=8gR_0S_0/U_0^2$) are listed, as well as the overloading parameter $\Delta q_s/q_{s0}$.

Table 1. Summary of numerical test cases.

Test	z_{b0} (m)	h_0 (m)	U_0 (m/s)	S_0	f_0	q_{s0} (m^3/s)	$\Delta q_s/q_{s0}$	Q_0 (m^3/s)
1	0.048	0.038	0.60	0.01122	0.066	$2.25\ 10^{-6}$	-1	0.0038
2	0.048	0.041	0.70	0.01122	0.051	$4.52\ 10^{-6}$	-1	0.0048
3	0.047	0.046	0.73	0.01122	0.051	$5.10\ 10^{-6}$	-1	0.0057

Once the following three conditions were reached: 1) equilibrium between erosion and deposition, 2) mobile bed stability, 3) uniform flow, then the second phase of the test could start.

Keeping constant the liquid discharge, sediment feeding was stopped and, simultaneously, elapsed time was measured with a chronometer. Erosion also started occurring in the mobile bed reach. At time t_1=10 minutes the run was interrupted and the chronometer stopped, by gradually decreasing the discharge. Once the channel was completely drained, the bed profile was acquired, by measuring the bed elevation, z_b, in given cross-sections, located at several abscissa x measured from the upstream end of the mobile bed reach. For each cross-section, the bed elevation was measured along five verticals and the average value was assumed as the cross-sectional bed elevation.

Then the run was started again, by gradually pumping the discharge into the channel, so that disturbances to the mobile bed could be minimised. Once the design value Q was reached, the chronometer was started again and at time t_2=20 minutes the run was interrupted and the chronometer stopped as explained before. Measurements of the bed profile were acquired with the above procedure, that was also repeated at time t_3=30 minutes.

Application of the DORA model to experimental cases

In Figs. 3.a to 3.c some of the comparisons between the DORA model results and experimental data are shown.

At the transition from fixed to mobile bed, a process similar to local excavation downstream of a weir occurs. The strong soil erosion is related to the dissipation of the turbulent kinetic energy of the upstream current, and, of course, can not be simulated by means of a formula obtained in equilibrium conditions. The local excavation causes an input of solid load in the downstream stretch of the channel much higher than the one estimated by Ackers' formula. To somehow estimate this solid load, the solid volume between the time level t_1 and t_2 and between the abscissas x_1 and x_2 has been measured from the experimental bed profiles; abscissa x_2 is the abscissa where significant change in bed profile slope appears. The longitudinal extension of the excavated volumes increases between time t_1 and t_2. In order to use a fixed computational grid point between the two time levels the first computational node has been located in an intermediate position. It has been hypothesized that the solid volume moves through the first upstream node of the computational grid by means of a solid load linearly changing according to the schemes of Figs. 4.a and 4.b.

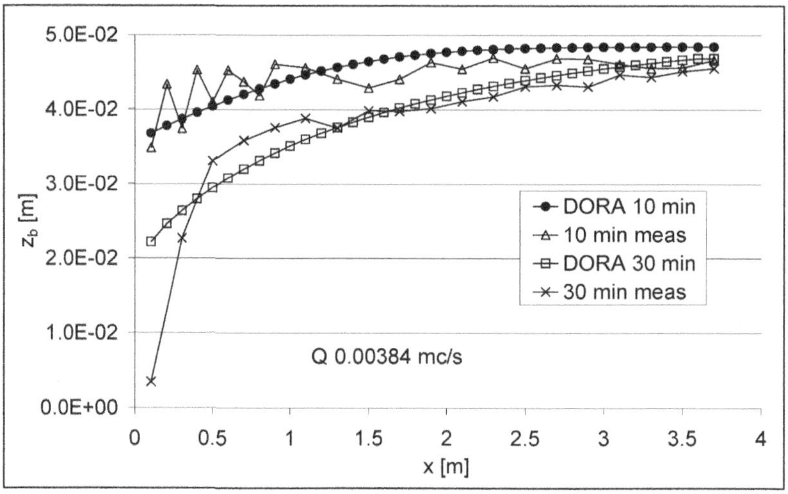

Figure 3.a Comparison between the DORA model results and measured data at 10 and 30 minutes (Q=0.00384 m^3/s).

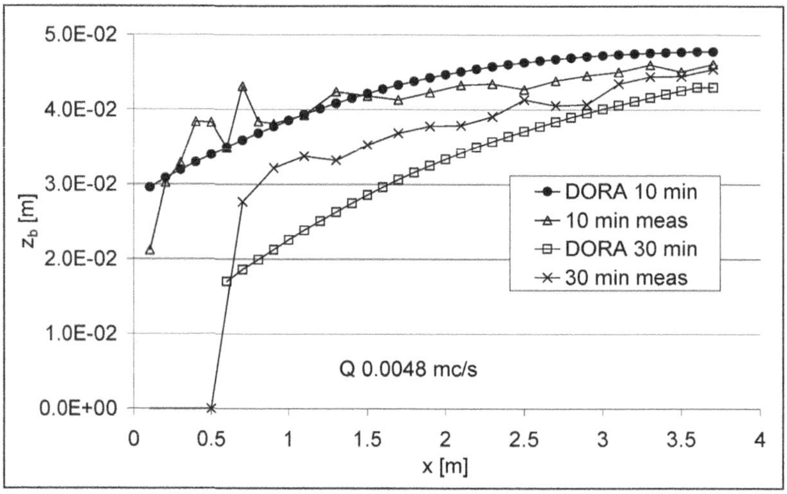

Figure 3.b Comparison between the DORA model results and measured data at 10 and 30 minutes (Q=0.0048 m^3/s).

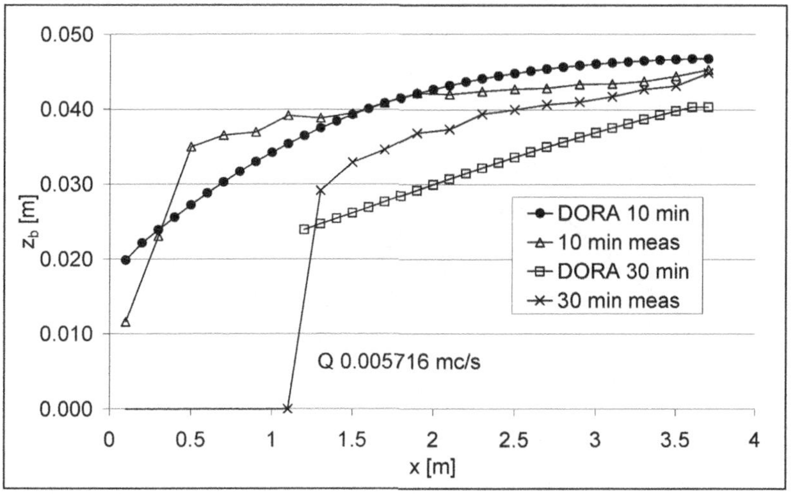

Figure 3.c Comparison between the DORA results and measured data at 10 and 30 minutes ($Q=0.0057\ m^3/s$).

The DORA model fits well experimental profiles especially for the lowest flow rate. In the cases of higher flow rates the agreement between the model and measured data is also good, especially for the duration of 10 minutes. For higher durations the DORA model overestimates the erosion in the considered stretch of the channel. It is very important to underline that no calibration of any parameter of the Ackers' formula has been carried out on the basis of the experimental data.

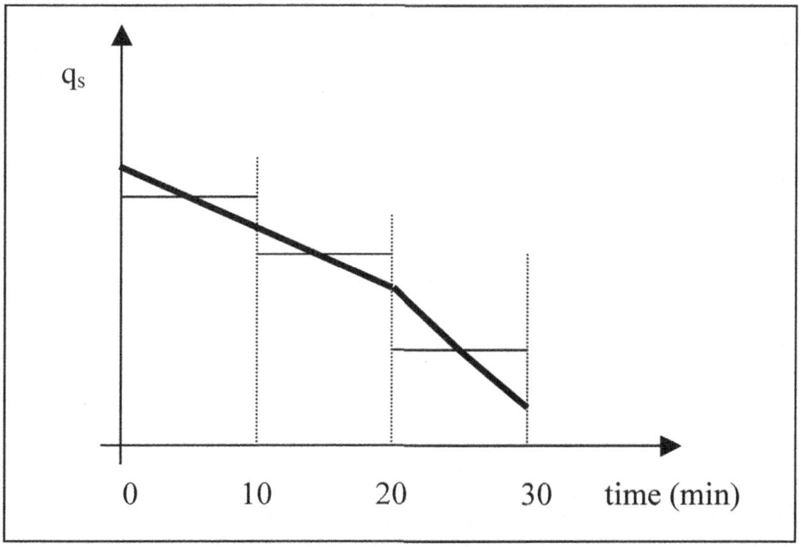

Figure 4.a Variation of bed load for tests 1 and 2.

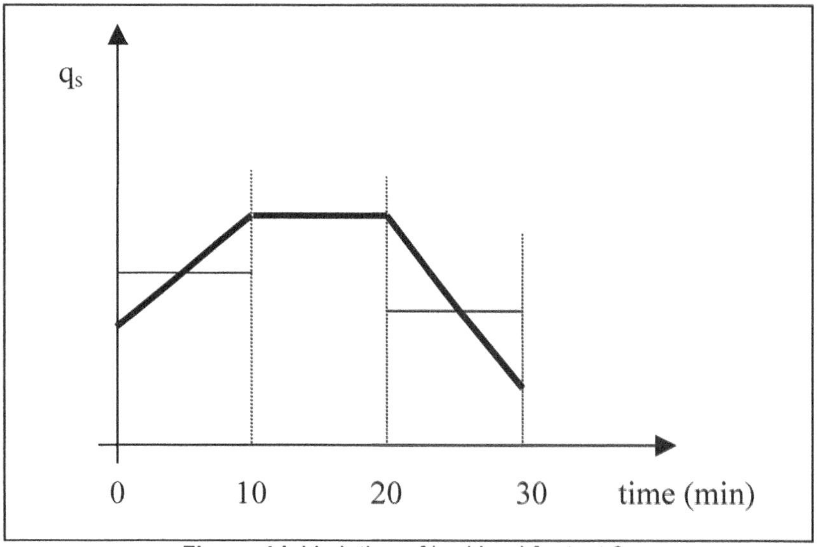

Figure 4.b Variation of bed load for test 3.

Conclusions

The DORA model has been applied for the numerical solution of the sediment transport problem adopting the diffusive hypothesis in the flow momentum equation. The model uses a fractional step procedure and has been tested in case of degradation, with data measured in a semi-ovoid perspex pipe. Sediment transport has been simulated by means of a formula tested in equilibrium condition; the local erosion at the transition from fixed to movable bed has been taken into account considering a solid load, computed on the basis of the measured bed profiles, entering the upstream node of the computational domain. It is important to underline that the model results are in quite good agreement with experimental data without any calibration of the parameters of the solid transport formula; on the other hand, in order to apply the model to field cases, an *ad hoc* investigation should be done to study local excavation transport processes, more likely to occur in real sewers.

Acknowledgements

The authors are grateful to Prof. F. Calomino and Prof. T. Tucciarelli for their help and valuable suggestions.

REFERENCES

Ackers, P. and White, W. R. (1973). Sediment transport: new approach and analysis. *J. Hydr. Div.* **99**, (11), 2041-2060.

Ackers P. (1984). Sediment transport in sewers and the design implications. *Proc. Int. Conf. "The Planning, Construction, Maintenance and Operation of Sewerage Systems"*, Cranfield BHRA, The Fluid Engineering Centre.

Aricò C. and Tucciarelli T. (2002) A partially coupled flow and transport model for sewer networks. *Proc. Int. Conf. "Sewer Operation and Maintenance"*, Bradford (UK), November 2002.

Aricò C. and Tucciarelli T. (2003). DORA model for flow and sediment transport simulation in sewer networks. Comparison with analytical solutions and experimental data. *Proc. "TCN-CAE 2003 Int. Conf. on CAE and Computational Technologies for Industry"*, October 2003.

Cao Z., Day R. and Egashira S. (2002). Coupled and decoupled numerical modelling of flow and morphological evolution in alluvial rivers. *J. Hydr. Eng.* **128**, (3), 306-321.

Correia L. R. P., Krishnappan B. G. and Graf W. H. (1992). Fully coupled unsteady mobile boundary flow model. *J. Hydr. Eng.* **118**, (3), 476-494.

Fischer H. B., List E. J., Koh R. C. Y. Imberger J., and Brooks N. H. (1979)."Mixing in Inland and Coastal Waters". Academic, New York, 104-138.

Kassem A. A. and Chaudhry M. H. (1998). Comparison of Coupled and Semicoupled Numerical Models for Alluvial Channels. *J. Hydr. Eng.* **124**, (8), 794-802.

Noto V. and Tucciarelli T. (2001). DORA Algorithm for network flow models with improved stability and convergence properties. *J. Hydr. Eng.* **127**, (5), 380-391.

Press, W. H., Flannery, B. P., Teukolsky, S. A., and Vetterling, W. T. (1988). "Numerical recipes", Cambridge University Press, Cambridge, U.K.

Tucciarelli T. and Fedele F. (2000). An efficient double order solution of the groundwater contaminant transport problem. *Proc. 13th Int. Conf. "Computational Methods in Water Resour."*.

Tucciarelli T. and Termini, D. (2000). Finite-element modeling of floodplain flows. *J. Hydr. Eng.*, **126**, (6), 416–424.

Advection and dispersion of solutes in sewer systems: verification of water quality models by tracer experiments

C.M.L. Flamink*, J.G. Langeveld and F.H.L.R. Clemens***

*Section of Sanitary Engineering, Faculty of Civil Engineering and Geosciences, Delft University of Technology (DUT), P.O. Box 5048, 2600 GA Delft, the Netherlands (C.M.L.flamink@citg.tudelft.nl; f.h.l.r.clemens@citg.tudelft.nl)

**Royal Haskoning, P.O. Box 151, 6500 AD Nijmegen, the Netherlands (j.langeveld@royalhaskoning.com)

Abstract Solute transport is one of the key processes with respect to predicting the soluble and fine suspended pollutant load discharged via combined sewer overflows (CSOs) and transported towards wastewater treatment plants (wwtps). This paper discusses the quality of available models for the transport of solutes in sewer systems. A number of tracer experiments and a 10 day measuring campaign provided the data to first check the validity of the well-known 1D advection –dispersion equation and subsequently the performance of two commercial software packages. Although commercial software packages show significant numerical dispersion, they can be used to properly model the transport of a soluble substance like ammonium during storm events, provided that the hydrodynamics are well known.

Keywords interactions within wastewater system, modelling, advection dispersion, tracer experiments

INTRODUCTION

Sewer flow quality modelling is one of the challenges within the field of urban drainage. The broad range of in-sewer processes, each having its own particular time- and space scale, and the inaccessibility of most of the sewer systems, hamper the confrontation of the simulation results of sewer models to measurements, as gathering reliable measurement data is extremely costly.

Nonetheless, a strong drive towards the use of sewer flow quality models can be noticed. The European Water Framework (WFD, 2000) is partly based on the widely accepted integrated, immission based approach, thereby urging scientists to deliver models able to simulate receiving water quality.

Recent research has shown that, with respect to the interactions within the wastewater system, modelling of dissolved compounds in sewer systems is necessary to be able to predict wastewater treatment plant performance (Langeveld *et al.*, 2003). Furthermore, the accurate modelling of the advection-dispersion process is a prerequisite for sewer suspended solids modelling, as the advection-dispersion modules should be able to model the steep gradients in concentration due to sudden re-erosion of sewer sediment.

The advection-dispersion process describes the movement of the pollutants within the aqueous phase through the sewer system. Typically, advection-dispersion involves the

 Sewer Networks and Processes within Urban Water Systems edited by J.-L. Bertrand-Krajewski, M. Almeida, J. Matos & S. Abdul-Talib. ISBN: 1 84339 506 1.

transport of dissolved and fine suspended substances within the wastewater, but also of bigger particles whilst eroded.

The advection part describes the carrying along of pollutants with the flow at the average flow velocity, whereas the dispersion describes to what extent the actual transport of pollutants deviates from the average flow velocity. Most available sewer models, e.g. MOUSE, Hydroworks and SOBEK, also provide modules for advection-dispersion [Bouteligier *et al.* (2001)], based on the general equation for the 1-dimensional description of the advection-dispersion process in sewer systems:

$$\frac{\partial c}{\partial t}+u\frac{\partial c}{\partial x}-K\frac{\partial^2 c}{\partial x^2}=0 \qquad \text{(eq. 1)}$$

where:

c	concentration	(kg/m^3)
u	flow velocity	(m/s)
t	time	(s)
x	space step	(m)
K	dispersion coefficient	(m^2/s)

However, the numerical implementation of the advection-dispersion equation, and therewith the simulation results, differs strongly. Furthermore, the simplification to a 1-dimensional description of the transport of pollutants within the aqueous phase has implications for the quality of the simulation results. During storm runoff manholes may surcharge resulting in a complex 3-dimensional flow field. Tracer experiments under both lab (Guymer and O'Brien (1995); Guymer et al. (1998)) and field (Boxall et al. (2003)) conditions showed the exchange of pollutants between the flow and the wastewater within the surcharged manhole to depend on both the flow and the level of surcharge. The concentration profiles measured in the experiments under surcharged conditions are often significantly skewed (i.e. having a long tail), where the advection dispersion equation can only predict spatial concentration profiles of Gaussian shape.

It could therefore be questioned to what extent the transport of solutes in sewer systems can be described by the 1D-advection-dispersion equation widely used in today's software packages. According to Bouteligier et al. (2001), today's software (in their case Infoworks, which models only advection, accounting implicitly for the physical dispersion by the numerical dispersion and the representation of the manholes as being completely mixed) is capable of properly describing the ammonium concentrations during wwf, provided that the hydrodynamic conditions are well-described and a reliable dwf profile is available. The successful application of Infoworks by Bouteligier et al. (2001) implies that either concentration gradients in sewer systems with respect to ammonium are small enough not to cause significant (numerical) problems or the 1D advection-dispersion equation as implemented in Infoworks is valid for solute transport in sewer systems, contrary to the findings of (Boxall et al. (2003)) and Guymer et al., 1996).

In this paper it will be analysed which of these two explanations does hold and to what extent it is possible to describe the transport of solutes in sewer systems with the 1D advection dispersion equation as implemented in today's commercial softwares.

Methods and materials

In order to determine whether advection-dispersion in sewers can be described with the 1-dimensional description, tracer experiments have been performed in two urban drainage systems. Furthermore, with the results of the tracer experiments the modelling of advection-dispersion by the commercial software packages has been verified under both dry and wet weather conditions.

Experiments

The tracer experiments consisted of dosing a NaCl solution (87.5 g/l) in a manhole (dosing manhole) and measuring conductivity (WTW Tetracon 325) every 5 seconds at 2 locations downstream (measuring manholes). The same sensor was used to measure upstream and downstream, as the travel time of the trace allowed transport of the sensor from upstream to downstream. It was assumed that the trace was fully mixed before arriving at the upstream measuring location. This was confirmed by measurements while moving the sensor along the cross section.

Tracer experiments have been carried out during dwf in 2 reaches with no side connections of the sewer system of Loenen (The Netherlands), see figure 1 and 2. During wwf a reach of the sewer system of Ulvenhout has been used, see figure 3. In both drainage systems, an extensive network of water level sensors and rain gauges has been installed to allow event-based calibration of the hydrodynamic model.

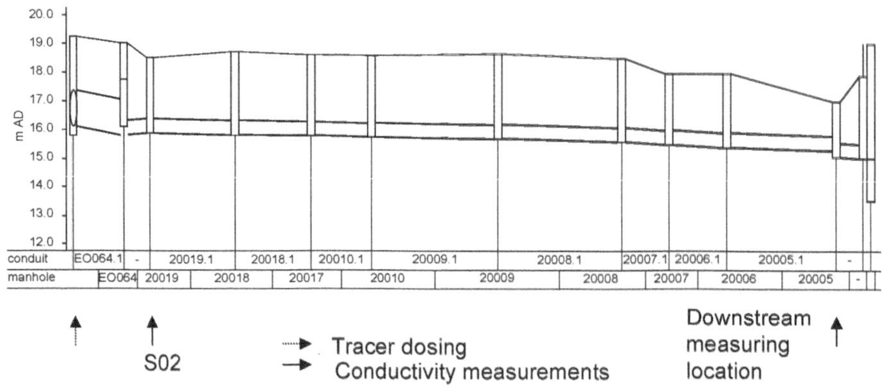

Figure 1 Loenen (dwf), 454 m reach, ∅ 0.5 m

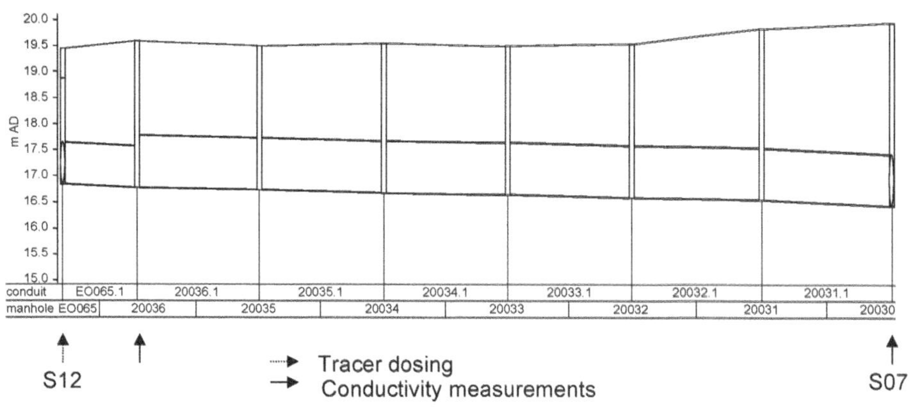

Figure 2 Loenen (dwf), 233 m reach, ∅ 1 m

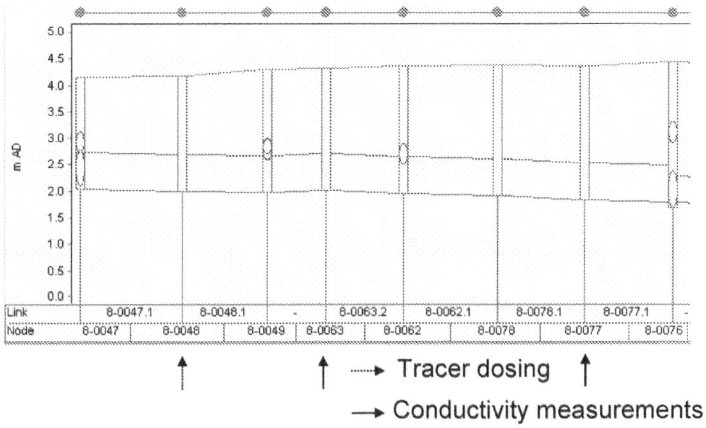

Figure 3 Ulvenhout (wwf), 142 m reach, ∅ 0.7 m

Apart from the tracer experiments a 10-day measuring campaign has been carried out in the sewer system of Ulvenhout. Samples have been taken as indicated in figure 3 and analysed for ammonium. During dry weather, a time resolution of one hour was used, which was lowered during storm events to 10 minutes. This enabled a more meticulous verification of the advection-dispersion modelling than often found in literature.

Modelling solute transport

Based on equation (1) an advection-dispersion model has been set up in Matlab. For this model a first order upwind scheme has been used in combination with the van Leer Limiter, see equation 2. Using this limiter, steep concentration gradients, due to a sudden increase of the concentration, can be modelled second order accurate as well (Vreugdenhil and Koren, 1993).

$$C = c_{i-1} + (1-\sigma)\frac{(c_i - c_{i-1})(c_{i+1} - c_i)}{(c_{i+1} - c_{i-1})} \qquad if \theta \geq 0$$

(eq. 2)

$$C = c_{i-1} \qquad if \theta < 0$$

where

σ	courant number	
c	concentration	(kg/m^3)

$$\theta = \frac{(c_i - c_{i-1})}{(c_{i+1} - c_i)} \qquad \text{(eq. 3)}$$

Figure 4 gives an example of the use of the Van Leer Scheme. The dotted line gives the inflow concentration, in this example at the upstream location of the sewer reach given in figure 1. The thick black line shows the simulation result at the downstream measuring location, without applying dispersion, showing that the numerical diffusion is negligible. The thick grey line shows the simulation result obtained with an ordinary first order upwind scheme, which introduces a significant amount of numerical dispersion.

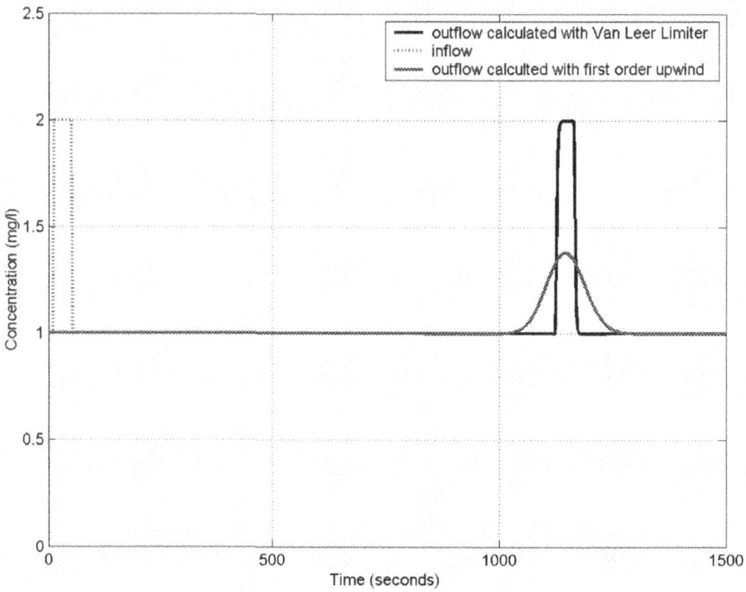

Figure 4 Test of numerical scheme modelled with Matlab

By comparing this model with the results of the tracer experiments it can be determined whether advection-dispersion in sewers can be described with the 1-dimensional description or more complex models as suggested by Guymer *et al.* (1996) are necessary.

In order to verify the modelling of advection-dispersion by the commercial software packages, the results of the tracer experiments have been compared to the results calculated with Hydroworks (Wallingford software 2000) and the simplest version of SOBEK (WLDelft 1998), the combination of the sewer flow and water quality module. Hydroworks only models advective transport, therefore the dispersion coefficient in SOBEK has been taken 0 in order to be able to make a good comparison. Huisman et al. (2000) showed MOUSE (using the Manning approach) to have a numerical dispersion larger than the actual physical dispersion. As a result, calibration of the dispersion coefficient, even under dwf conditions, proved to be impossible.

In this research the ammonium concentrations during a storm event in the 10-day measuring campaign have been simulated with Hydroworks and compared to the measured concentrations.

Results and discussion

1D advection-dispersion equation in sewers

Figure 5 gives the results of the tracer experiments carried out during dwf in Loenen. The upper graphs show both the measured conductivity at the upstream and downstream manhole and the modelled conductivity downstream. The lower graphs show the residuals after manually fitting the model to the data by adjusting the dispersion. Although the results could be improved by further calibration, it is clear that the advection-dispersion process can be modelled properly by adjusting the dispersion coefficient. In these cases values have been found for the dispersion coefficient of respectively 0.15 ± 0.002 m^2/s for the 454 m long sewer reach with a diameter of 0.5 m and 0.04 ± 0.0001 m^2/s for the 233 m long, 1 m diameter sewer reach. The dispersion coefficients have the same order of magnitude as values found under similar (dwf) conditions. Huisman *et al.* (2000) reported a dispersion coefficient of 0.05 m^2/s and Boxall *et al.* (2003) reported a mean dispersion coefficient of 0.06 m^2/s with a standard deviation of 0.05 m^2/s.

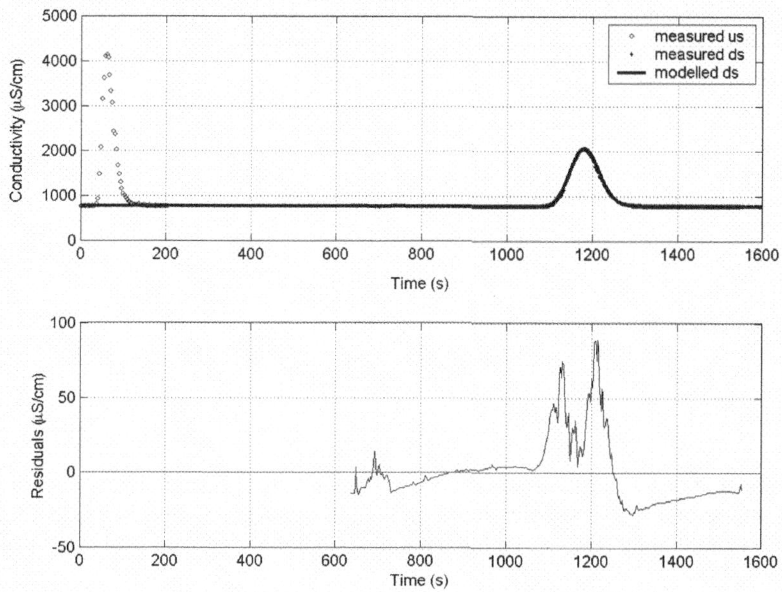

Figure 5 Results Matlab model. 454 m sewer reach, diameter 0.5 m, shown in figure 1. The model results and measurements are in accordance. The long 'tail' of the residuals after t = 1250 s illustrates that the transport through the sewer is not fully 1-dimensional.

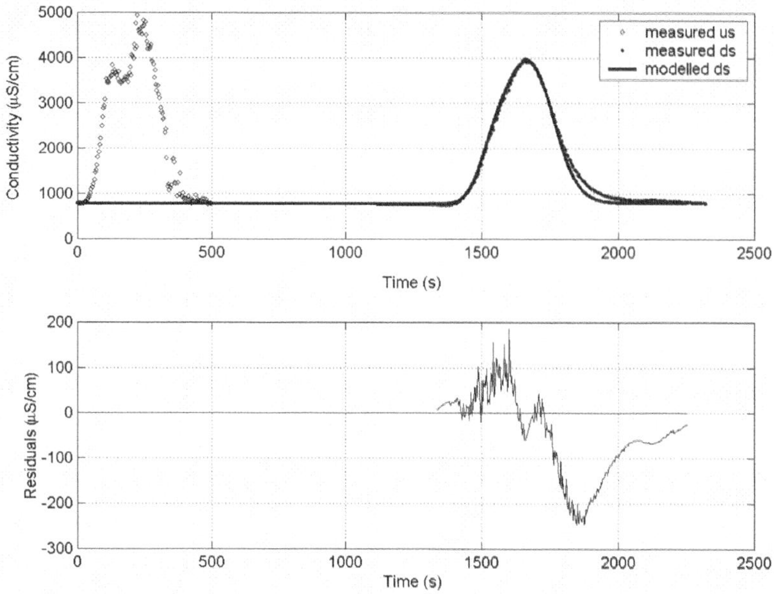

Figure 6 Results Matlab model. 233 m sewer reach, diameter 1 m, shown in figure 2. the somewhat strange shape of the conductivity curve measured upstream is due to the fact that two intermittent doses of trace started to overlap. At the downstream location these two peaks are not noticeable anymore.

Under wwf conditions, the sewer system became surcharged and showed a backwater effect. The NaCl solution sank to the bottom and due to the low turbulence didn't fully mix. This phenomenon was observed by moving the sensor up and down in the sewage. Below a depth of approximately 10 cm the conductivity measured was over 10.000 μS/cm, whereas only 5 centimeters above this layer the conductivity was 300 μS/cm.

Results Hydroworks and SOBEK

The results calculated by Hydroworks and SOBEK for the 454 m reach are given in figure 7. Both models are not able to reproduce the measured concentration. Hydroworks describes the moment the peak passes reasonably, but underestimates the maximum value. This is partly due to the fact that the minimal timestep in the input as well as the output in Hydroworks is 15 seconds and partly due to numerical diffusion. The moment on which the peak passes is determined by the flow velocity in the model. Since the flow velocity depends on the flow input and this is entered in m^3/s with maximal 3 decimals, it was not possible to reproduce the moment of the peak exactly. In the Matlab simulations, the flow velocity was derived from the measured travel time of the tracer. As the experiments lasted only 10-15 minutes, the changes in dwf could be neglected. The velocity in the first experiment in the sewer reach given in figure 1 was 0.41 m/s and the flow rate 14 l/s. This flowrate was confirmed by flow recordings in the pumping station directly downstream of this sewer reach.

In the results obtained with SOBEK a peak is hardly recognisable. Whereas SOBEK does not have the aforementioned limitations of Hydroworks, the number of decimals of the input of the flow is sufficient and the adjustable minimal timestep is 1 second, the bottleneck in this case is the fact that the spacestep by default equals the length of the reach. The result is a very large numerical dipersion.

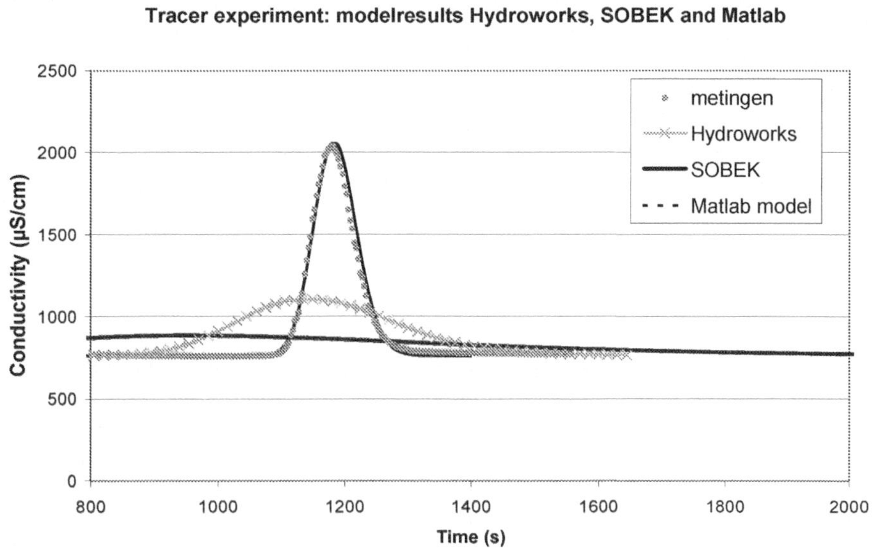

Figure 7 Results Hydroworks, SOBEK and Matlab

Figure 8 gives the results of simulations with the second available version in SOBEK, the combination of the modules sewer flow (module also applied in figure 7), channel flow (an add-on module allowing to adjust the space step for the water quality simulations) and water quality. Hereby, two numerical schemes, 'Fully implicit iterative method', which is an ordinary first order upwind like the one shown in figure 4 and 'Modified flux corrected transport', which is a numerical scheme correcting for numerical dispersion like the Van Leer Scheme, have been calculated with different spacesteps.

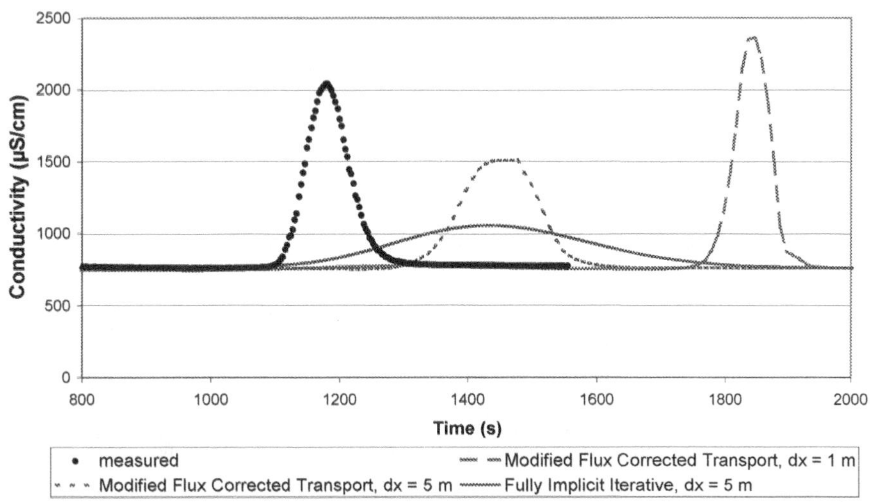

Figure 8 Results SOBEK (modules sewer flow, channel flow and water quality)

Compared to the basic version the calculated peak values match the measured values much better, but the travel time doesn't fit the measured travel time. In figure 8 it can be seen that the simulated peaks arrive later than the measured peaks and also that this delay increases with a decreasing space step. This was caused by the fact that in the model the calculation nodes applied have a certain dimension. By decreasing the space step, the number of calculation nodes increases, i.e. at each calculation grid point a calculation node is introduced. As these gridpoints were not dimensionless, this approach effectively increases the length of the sewer reach. With a space step of 1 meter this results for this reach in an extra delay of approximately 7 minutes, see figure 8. Converted to a complete sewer system, where transport distances amount up to 2 km, this gives an error of almost half an hour, which can lead to significant errors for e.g. RTC applications.

Ammonium concentration

The simulated and measured ammonium concentration show the same trend, see figure 9, which is in line with the conclusions of Bouteligier et al. (2001). This indicates that the concentration gradients in sewer systems with respect to ammonium are small enough not to cause significant (numerical) problems. However, the model systematically underestimates the ammonium concentration during the storm event. A possible explanation for this is that the ammonium concentration in the rainwater is, as opposed to assumed in the model, not zero but likely approximately 1 mg/l (Stolk, 2001) or even higher (Bertrand-Krajewski et al., 1995). Another explanation can be that ammonification decreases during wwf.

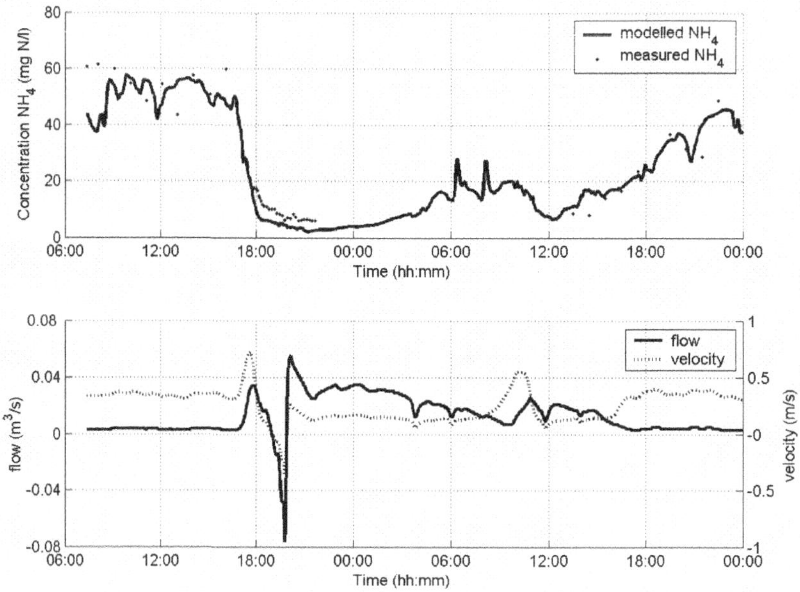

Figure 9 Top: Measured and simulated ammonium concentration, bottom: flow rate and velocity during the experiments

Conclusions

The tracer experiments provided valuable data during dwf. During wwf special attention is necessary to check whether the dosed tracer completely mixes over the cross section of the sewer. The results of the tracer experiments and the measuring campaign revealed that:

- Advection-dispersion in sewers can, according to the results of the dwf tracer experiments, be described with the general 1D advection dispersion equation.
- Hydroworks and SOBEK are not capable of reproducing the results of the tracer experiments due to a too large numerical dispersion, which has been emphasised by the high gradients in the model input.
- The ammonium concentration during wwf can be modelled accurate enough, given the overall uncertainties in modelling of sewer systems and wastewater treatment plants, to be used to predict wastewater treatment plant loading. As the ammonium concentration in the wastewater does not show high gradients, the effect of the numerical dispersion does not severely affect the quality of the simulation results, like with the tracer experiments.

Acknowledgements

The first author gratefully acknowledges the funding by the partners of the working group 'Interactions within wastewater systems II': Arcadis, DWR, Grontmij, Groot Salland, Hollands Noorderkwartier, Royal Haskoning, Tauw, De Dommel and Witteveen+Bos

References

Bertrand-Krajewski, J.-L., Lefebre, M, Lefai, B. and Audic, J.-M. (1995). Flow and pollutant measurements in a combined sewer system to operate a wastewater treatment plant and its storage tank during storm events. *Wat.Sci.Tech.* **31**(7), pp. 1-12

Bouteligier, R., Vaes, G., Berlamont, J., Assel, J. van and Gordon, D. (2001). Water Quality Model set-up and Calibrating – a case study. WaPUG Autumn Conference, 14-16th November 2001 – Blackpool, UK (10 pages)

Boxall, J., Shepherd, W., Guymer, I. and Fox, K. (2003). Changes in water quality parameters due to in-sewer processes. *Wat. Sci. Tech.*, 47(7-8), 343-350.

Guymer, I. and O'Brien, R. (1995). The effects of surcharged manholes on the travel time and dispersion of solutes in sewer systems. *Wat.Sci.Tech.*, **31**(7), 51-59.

Guymer, I., O'Brien, R. and Harrison, C. (1996). Representation of solute transport and mixing within a surcharged benched manhole using an aggregated dead zone (ADZ) technique. *Wat Sci.Tech.*, **34**(3-4), 95-101.

Guymer, I., O'Brien, R., Mark, O. and Dennis, P. (1998). An investigation of fine sediment mixing within free-flowing and surcharged manholes. *Wat.Sci.Tech.* **37**(1), pp 215-222

Huisman J.L., Burckhardt S., Larsen T.A., Krebs P., and Gujer W. (2000). Propagation of waves and dissolved compounds in a sewer. *ASCE J.Env. Eng.*, **126**(1), 12-20.

Langeveld, J.G., Clemens, F.H.L.R. and Graaf, J.H.J.M. van der. (2003). Interactions within the wastewater system: requirements for sewer processes modelling. *Wat Sci Tech.*, **47**(4), 101-108.

Stolk, A.P. (2001). *Landelijk Meetnet Regenwatersamenstelling. Meetresultaten 2000.* RIVM Rapport 723101 057/ 2001 Dutch National Precipitation Chemistry Network. Monitoring results for 2000 (in Dutch), 67 pages

Vreugdenhil, C.B. and Koren, B.(1993) Numerical methods for advection diffusion problems. Braunschweig, Wiesbaden: Vieweg. 373 pages

Wallingford Software (2000). *HydroWorks Documentation.* On-line documentation. Version 6.0. Wallingford Software Ltd.

WFD (2000). *The EU Water Framework Directive- integrated river basin management for Europe*

WLDelft (1998). *SOBEK Manual.* SOBEK Urban, version 2.04. May 1998. WL Delft Hydraulics

Measuring Oxygen Transfer in Sewer Drops

A. Soares*, M. C. Almeida, J. Matos*****

*Trainee Research Assistant, Instituto Superior Técnico, Technical University of Lisbon, Av. Rovisco Pais, 1049-001, Lisbon, Portugal *(anaalmsoares@hotmail.com)*
**Research Officer, National Laboratory of Civil Engineering, Hydraulics and Environment Department, Av. do Brasil, 101, 1700-066, Lisbon, Portugal (*mcalmeida@lnec.pt*)
***Assistant Professor, Department of Civil Engineering and Architecture, Instituto Superior Técnico, Av. Rovisco Pais, 1049-001, Lisbon, Portugal, (*jm@civil.ist.utl.pt*)

Abstract An experimental study has been carried out in order to deepen the knowledge on the reaeration performance of sewer drops, by obtaining additional field measurements in the partially separate sewer at Estoril Coast interceptor system. New experimental data were obtained on simple vertical free-fall drops, sloping ramp drops and backdrops, for fall heights ranging from 0.20 to 1.45 m, flow rates from 17 to 350 l/s and temperature varying between 20 and 21.5 degrees. Both hydraulic and water quality parameters were measured in the vicinity of the drop. In this paper, the overall results of field measurements are presented and the existing formulae for predicting the reaeration in sewer drops are reviewed in light of the data obtained in prototypes. Some important aspects related with the accuracy of the measurements are outlined and suggestions are provided for further developments.

Keywords Sewer drops, reaeration, deficit ratio

INTRODUTION

Biodegradation processes occur during transport of wastewater through sewer systems, consuming the oxygen dissolved in the wastewater. Despite the continuous reaeration through the air liquid interface, consumption due to biochemical processes is often higher, resulting in the decrease of the dissolved oxygen concentration in the flow direction.

Air entrainment occurring at singularities (e.g. drops, curves and junctions) provides air bubble aeration, contributing significantly to the increase of the dissolved oxygen content. Even though the beneficial role of sewer drops in the increase of dissolved oxygen is well known, the number of prototype studies dealing with raw wastewater is very limited.

Most studies described in the literature on reaeration of water by structures that create high turbulence are related to river water quality (e.g. Gameson *et al.*; 1958, Avery and Novak, 1978; Nakasone, 1987; Labocha *et al.*, 1996; Watson *et al.*, 1998). Studies are mostly on aeration at weirs and cascades and experiments have often been done in laboratory facilities using either clean water, polluted river water or treated effluents. To date, studies which have been conducted in sewer drops are very limited, examples being those carried out by Matos (1991) on simple vertical free fall drops in a small diameter sewer, for low flows, and those by Almeida (1999), on simple vertical free-fall drops, sloping ramp drops and backdrops in a foul sewer at Estoril Coast interceptor system, Portugal.

The overall oxygen transfer at hydraulic structures such as drops can be measured by the deficit ratio r defined by

 Sewer Networks and Processes within Urban Water Systems edited by J.-L. Bertrand-Krajewski, M. Almeida, J. Matos & S. Abdul-Talib. ISBN: 1 84339 506 1.

$$r = \frac{C_s - C_u}{C_s - C_d} \quad (1)$$

where C_s is the dissolved oxygen concentration at saturation and C_u and C_d are the upstream and downstream dissolved oxygen concentrations at the drop, respectively.

Matos (1991) performed measurements at prototype vertical free-fall drops in small diameter sewers (diameter of 0.2 m), with drop height ranging from 0.10 to 1.75 m, temperature varying between 18 and 19.5°C and flow rates not exceeding, usually, 1 l/s, tailwater depths of 0.1 m and proposed the following expressions for the deficit ratio.

$$r = e^{0.29H} \quad (2)$$

$$r = \frac{1}{e^{0.125H^2 - 0.45H}} \quad (3)$$

where H is the head loss at the drop, i.e., the difference between the energy line upstream and downstream of the drop.

More recently Almeida (1999) conducted prototype measurements on a foul sewer at Estoril Coast interceptor system, Portugal. Measurements were carried out at 15 drops, with fall height ranging from 0.08 to 1.62 m, temperature varying between 19 and 22°C, and dry weather flow rates ranging from 150 to 330 l/s. Several types of drops were analysed, namely vertical free-fall drops, sloping ramp drops and backdrops. Almeida (1999) has analysed the results by subdividing the drops into two groups by type of structure, one (i) comprising the sloping ramps and small curves presenting smooth flow, without any other associated feature, and the other (ii) including the sloping ramps followed by bends larger than 30°, free-fall drops and backdrops.

For the overall prototype drops where significant turbulence was created (group (ii)) two expressions were obtained, one linear and other exponential that provide reasonable fit to the data (Almeida, 1999, Almeida *et. al.*, 1999). In both equations d is the drop height in meters.

$$r = e^{0.24d} \quad (4)$$

$$r = 0.28d + 1 \quad (5)$$

Even though the drop height has been recognized as the most important variable influencing the deficit ratio in each type of drops, the impact of the flow rate and the tailwater depth (or flow depth at the toe of the drop) might also be of importance. In most cases of prototype sewers, the flow depth at the toe of the drop is small. The water quality is also a factor that can influence the reaeration but existing expressions, derived from studies in rivers or in sewers, do not incorporate this factor in a consistent way.

The objectives of the present study are: carry out field measurements on oxygen transfer at drops in sewers; study the effects of oxygen transfer resulting from different types of drops and establish an expression based on observed results; and compare the proposed expression to existing expressions.

EXPERIMENTAL WORK AND METHODOLOGY

A new set of measurement campaigns has been carried out based on the procedure suggested by Almeida (1999), as follows:

a) 44 drops were investigated from which nine have been selected based on local conditions, minimum drop height (0.22 m), minimum pipe diameter (500 mm), and appropriate conditions for installation of the equipment and executing the work under safe conditions;
b) In each drop, two sewers were selected to carry out the measurements of dissolved oxygen concentration, in accordance to Figure 1:
   i. control sewer: to assess the unit oxygen balance under undisturbed conditions;
   ii. sewer with the drop: to assess the contribution of the drop to dissolved oxygen;
c) In the control sewer, the upstream and downstream dissolved oxygen concentration was measured at the boundary manholes, to estimate the balance between the consumption and reaeration through the free-surface per unit time;
d) In the sewer length containing the drop, the dissolved oxygen concentration was measured at the manholes upstream and downstream of the manhole where the drop is located, to estimate the increase of dissolved oxygen in the drop, also making use of the balance between the consumption and reaeration through the free-surface obtained in the control sewer (c).

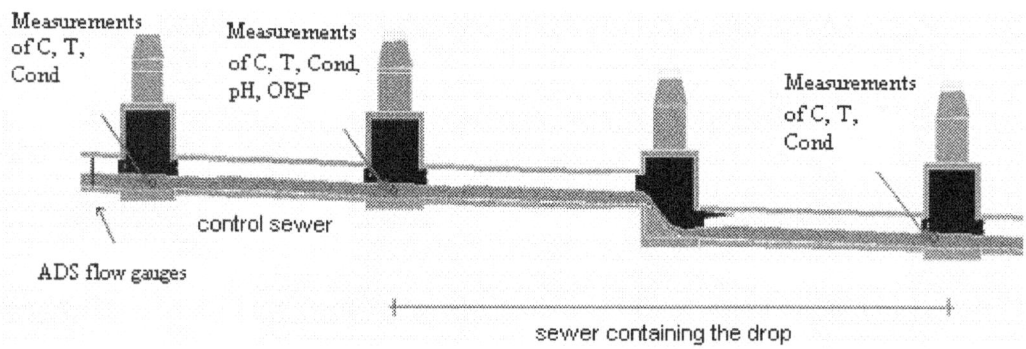

C – dissolved oxygen concentration; T – temperature; Cond – conductivity and ORP – redox potencial

Figure 1 – Scheme of measurements carried out.

The measurements of dissolved oxygen concentration were obtained by using two multiparametric sensors (YSI 556 MPS). Additionally these sensors provided information on temperature, conductivity, pH and redox potential.

Measurement of flow depth, velocity and discharge were conducted using ADS flow gauges previously installed, equipped with three sensors: one quadripole ultrasonic sensor for level measurement located in the pipe soffit, a pressure sensor for level measurement located in the pipe invert, and a Doppler effect velocity sensor also located in the pipe invert. The sensors were installed in a ring and connected to a datalogger by individual cables.

Furthermore, in each drop, samples were collected for the determination of total and dissolved COD, BOD_5, total, volatile and suspended solids, ammonia, nitrates and oxygen uptake rate. Continuous measurement of the oxygen uptake rate was carried out by using a custom built respirometer (Almeida, 1999).

DISCUSSION OF THE RESULTS

During each measurement campaign, discharge, velocity and flow depth showed small variations, since the duration of a campaign was relatively short (about four hours) and variations on dry weather flows are relatively slow (Figure 2).

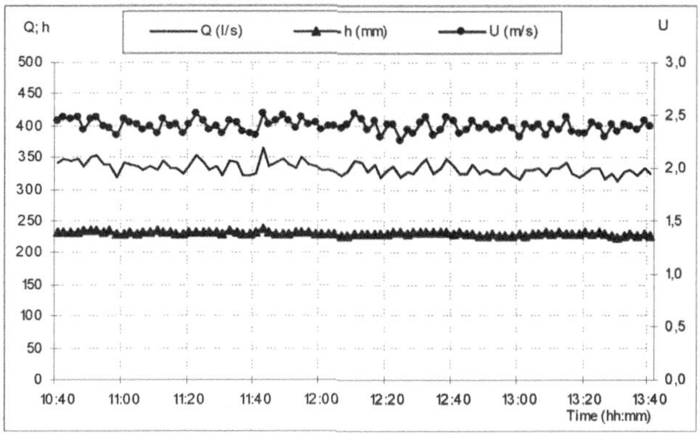

Figure 2 – Discharge, flow depth and velocity versus time, for a given experimental run.

Flow measurements in foul sewers are not straightforward, problems depending on local characteristics, flow type, wastewater characteristics and environmental conditions present in confined spaces. In the present study some problems were detected by checking the flow versus the flow depth scattergraphs (Figure 3).

Where problems were detected additional verifications were carried out and, in some cases, the ring had to be moved to a different location. Despite of the problems encountered it was possible to estimate the flow in the various locations.

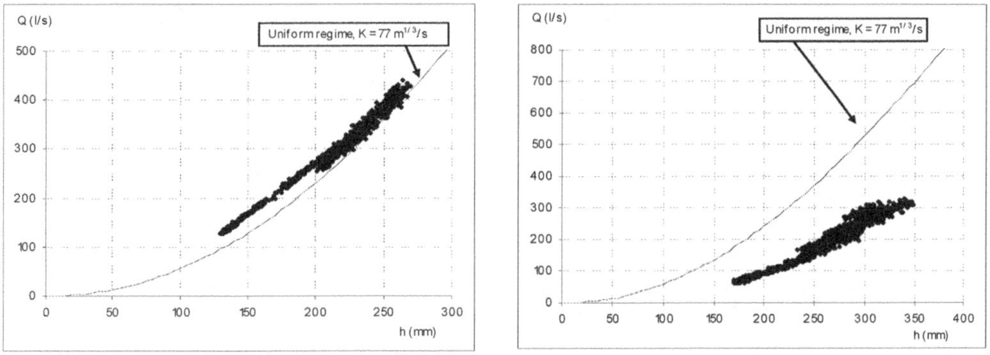

(a) approaching the uniform regime (b) away from the uniform regime

Figure 3 – Discharge versus flow depth, for a given experimental run.

Discharge measurements should be conducted in sections where the quasi-uniform regime is established. However, in real sewers, this situation is extremely hard to find. Furthermore, factors such as suspended solids variation and the presence of particles in sewers atmosphere, vibrations and noise can affect ultrasonic velocity sensor and the ultrasonic level sensor, respectively.

Measurements using the multiparametric sensors *in situ* were mainly affected by the difficulties in stabilizing the sensors when the flow was particularly turbulent leading to instability in the oxygen measurement as result of emersion of the sensor. Measurement of oxygen was specially affected in this situation resulting in a number of measurements that could not be used for the study. An example of the data obtained is given in Figure 4.1 and Figure 4.2. The complete data set for all drops investigated can be found in Soares (2003).

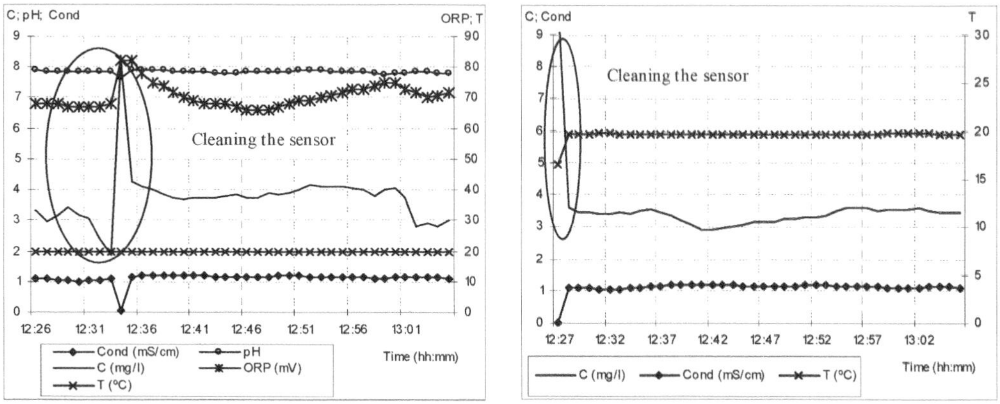

(a) upstream section (b) downstream section

Figure 4.1 - Results of measurements by multiparametric sensors installed in boundary sections of control length (06-06-2002).

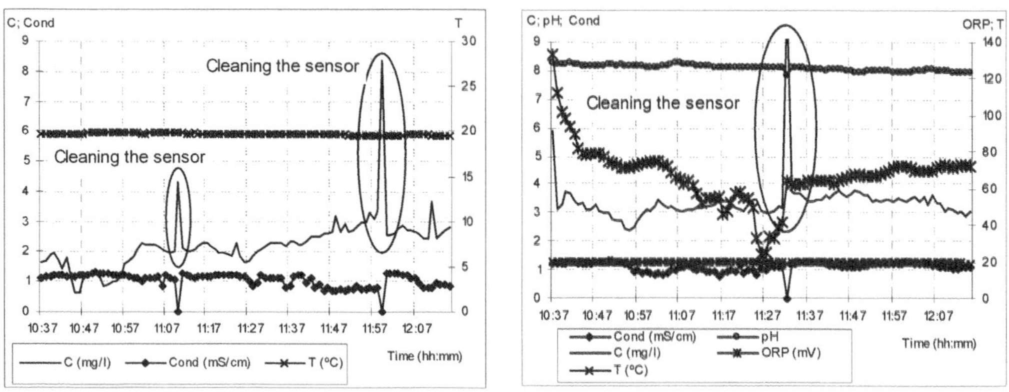

(a) upstream section (b) downstream section

Figure 4.2 - Results of measurements by multiparametric sensors installed in boundary sections of length containing the drop (06-06-2002).

Even when measurements are planned specifically to the purpose of quantifying the effect of turbulence in the reaeration of the wastewater, the hard conditions in existing sewers make it difficult to obtain data of good quality and to carry out a large number of measurements. Any immerse sensor in wastewater is subject to chemical attach and mechanic erosion, accumulation of sediments, grease and different solids.

Regarding dissolved oxygen measurements problems include: damaging of the sensor membrane; sensor calibration often affected during measurements; obstruction of the probe due to the attachment of detritus characteristic of the domestic wastewater (toilet paper, plastic, grease and oil, among others) and emersion of the probe due to the turbulent flow.

The continuous periods of measurements are usually short since cleaning the sensors was often necessary due to the accumulation of detritus and sticking of oil and grease.

Analytical work on samples was carried at LNEC's laboratory using standard methods. Analysis of results did not provide an acceptable relation between the different parameters and the value of the dissolved oxygen deficit ratio. Only the values of the oxygen uptake rate are more related to this difference as it could be expected from the oxygen balance.

The variation in the concentration of dissolved oxygen from upstream to downstream is a result of a simple oxygen balance in flowing wastewater in upstream length of the drop and downstream length of the drop, given by

$$\frac{dC}{dt} = r_c - (r_w + r_b) \tag{6}$$

where C is the dissolved oxygen concentration (mg/l), r_c – reaeration rate through the liquid surface (mg/(lh)), r_w – rate of consumption of dissolved oxygen within the bulk water (mg/(lh)) and r_b – rate of consumption of dissolved oxygen due to the slime layer(mg/(lh)).

The balance between microbial consumption processes (within the bulk water or slime layer) and reaeration along the reach, by unit time, r_c – (r_w + r_b), was estimated by measurements of dissolved oxygen concentration in the control sewer and compared to the OUR obtained in the laboratory to detect eventual measurement errors.

The relation proposed by Gulliver *et al.* (1993) was used herein to adjust the deficit ratio r to that expected at 20°C

$$r_{20} = r_T^{\ 1/\alpha_T} \tag{7}$$

where r is deficit ratio to temperature 20°C, r_T deficit ratio to the given temperature and α_T empirical coefficient, given by

$$\alpha_T = 1.0 + 0.02103\,(T-20) + 8.261\times10^{-5}(T-20)^2 \tag{8}$$

The experimental data on deficit ratio r obtained in the present study are shown in Figure 5, as a function of the drop height. The data presented in Figure 5 was obtained for drops with height varying between 0.22 and 1.45 m and discharges between 16 and 350 l/s. Also included in Figure 5 are a selected set of data obtained by Matos (1991), and the data collected by Almeida (1999), both readjusted herein for the temperature of 20°C. Additionally, are represented equations (3) and (5). The expression to explain Figure 5 is equation (9), because it was found to provide a reasonable fit to all data, even though considerable scatter can be noticed.

$$r = 1 + 0.31d^{0.98} \qquad R^2 = 0.83 \tag{9}$$

In Figure 5, the data marked with circles were not considered in equation (9) because the measurements of dissolved oxygen were found to be inaccurate. Such data inaccuracy were likely caused by the following occurrences:

- *Data symbols* (i) *and* (iv) - they are relative to a situation where a sloping ramp drop was found immediately upstream of the vertical free-fall drop. For the calculation of the deficit ratio, it was assumed that the reaeration was directly proportional to the drop height, regardless of the type of drop, which is an oversimplification.

- *Data symbol* (ii) – correspond to a vertical free-fall drop followed by a large bend (of about 90º). The increase of the turbulence in the flow due to the bend was observed; hence the dissolved oxygen concentration should be larger than that expected on a similar drop but without the bend.
- *Data symbol* (iii) – in this case, the length of the sewer immediately downstream of the drop was small (L ≈ 5 m). The reaeration at the drop may therefore be underestimated since the complete dissolution of air bubbles was not likely to occur.
- *Data symbol* (v) – here the multiparametric sensors were installed on a floater, which was found to be disturbing the flow conditions in the vicinity of the probe, leading to flow separation and thus to the increase of dissolved oxygen content.

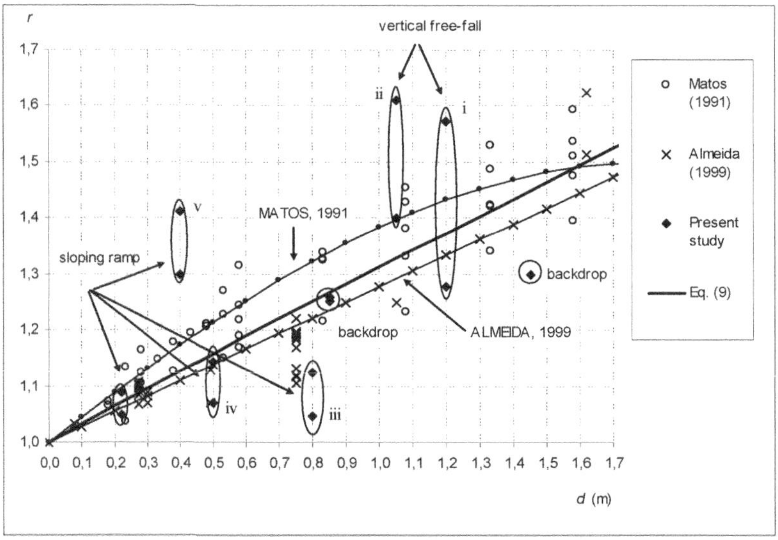

Figure 5 – Dissolved oxygen deficit ratio versus drop height.

The following observations can be drawn from Figure 5:

- The influence of drop height *d* on deficit ratio *r* is clearly shown.
- For identical drop height, the prototype data presented by Matos (1991) and by Almeida (1999) are in general not substantially different from those obtained in the present study, even tough the scatter of the data is noticeable.
- Structures creating more turbulence result in a significantly larger reaeration.
- The empirical regression equation provides values fairly similar to those given by Almeida (1999).
- The differences on the large-scale prototype data compared to those obtained by Matos (1991) may result from different type of drops, flow conditions (flows rates not exceeded 1 l/s, in the latter situation) and sewer diameters (diameter of 200 mm, in the latter situation).

Additionally, it has been observed that the values of the deficit ratio are extremely sensitive to errors in the dissolved oxygen concentration, and consequently field measurements are prone to a large scatter.

In the approach used for the calculation of the deficit ratio it is assumed that the balance between the total consumption and the continuous reaeration by the free surface, by time unity, remains constant on the different measurement pipes. This simplification can result in an additional error, which is difficult to estimate.

CONCLUSION

Measurements have been conducted in vertical free-fall drops, sloping drops and backdrops in prototype foul sewers. The data obtained and the empirical regression equations are similar to those obtained on prototype drops by Matos (1991) and Almeida (1999).

As already seen in a large number of studies, the height of the drop is the most important factor for the reaeration. The type of structure in terms of turbulence created is also relevant: drops in sloping ramps present smaller values of deficit ratio when compared to vertical free-fall drops and backdrops. Data was insufficient to establish the influence of discharge and water quality, measured by a number of usual parameters, on the deficit ratio value.

In order to deepen the knowledge on the hydraulics and oxygen transfer in sewer drops and hence to overcome the problems encountered in the field, a parallel study has been carried out in a hydraulic recirculating flume at the Laboratory of Hydraulics of the Technical University of Lisbon, IST. To date, results were gathered for vertical drops (Sousa *et al.* 2002, 2003, Soares, 2003). Further research includes additional laboratory experiments on other types of sewer drops (e.g., stepped cascades and backdrops).

ACKNOWLEDGEMENTS

The authors acknowledge the support of the Fundação para a Ciência e Tecnologia, Portugal, via the research project POCTI/36530/ECM/2000.

REFERENCES

Almeida, M C. (1999). Pollutant transformation processes in sewers under aerobic dry weather flow conditions. Ph.D thesis, Imperial College, London.

Almeida, M. C., Butler, D. and Matos, J. S. (1999). Reaeration by sewer drops. *Proc. 8th. Int. Conf. on Urban Storm Drainage*, Sydney, Australia, 413-420.

Avery, S. T. and Novak, P. (1978). Oxygen Transfer at Hydraulic Structures. *Journal of the Hydraulics Division*, ASCE, **104**(11), 1521-1540.

Gameson, A. L. H., Vandyke, K. G. and Ogden, C. G. (1958). The effect of Temperature on Aeration at Weirs. *Water and Water Engineering*, Nov., 489-492.

Gulliver, J. S. and Rindels, A. J. (1993). Measurements of Air - Water Oxygen Transfer at Hydraulic Structures. *Journal of Hydraulic Engineering*, ASCE, **119**(3), 327-349.

Labocha, M., Corsi, R. L. and Zytner, R. G. (1996). Parameters influencing oxygen uptake at clarifier weirs. *Water Environmental Research*, **68**(6), 988-994.

Matos, J. S. (1991). Aerobiose e Septicidade em Sistemas de Drenagem de Águas Residuais. Ph.D. thesis, IST, Lisbon (in portuguese).

Nakasone, H. (1987). Study of Aeration at Weirs and Cascades. *Journal of Environmental Engineering*, ASCE, **113**(1), 64-81.

Soares, A. A. (2003. Rearejamento em Quedas em Colectores de Águas Residuais. M.Sc. thesis, FCTUC, Coimbra (in portuguese).

Sousa, C. M. and Lopes, R. R. (2002). Hidráulica e Rearejamento em Quedas Verticais em Colectores. Research Report, IST, Lisbon (in portuguese).

Sousa, C. M., Lopes, R. R., Matos, J. and Almeida, M. C. (2003). Reaeration by Vertical Free-Fall Drops in Circular Channels. A Model Study". *Proc. XXX IAHR Congress*, Theme B, Thessaloniki, Greece, 361-367.

Watson, C. C., Walters, R. W. and Hogan, S. A. (1998). Aeration Performance of Low Drop Weirs. *Journal of Hydraulic Engineering*, ASCE, **124**(1), 65-71.

Electron Acceptor Utilization Rate in Municipal Wastewater under Anoxic Condition

Siti Maizurah Misuan and Suhaimi Abdul-Talib

Faculty of Civil Engineering, University Technology MARA, 40450 Shah Alam, Malaysia

Abstract Electron acceptor, electron donor and heterotrophic biomass are elements that must be present for microbial processes in sewer system to occur. Microbial processes in sewer networks may occur under aerobic, anoxic, or anaerobic conditions depending on the type of electron acceptor present. In the presence of dissolved oxygen (DO) aerobic process will occur. In the absence of DO, nitrate/nitrite will be used as electron acceptor, creating an anoxic condition. Finally, if DO, nitrate and nitrite are depleted from the system, then anaerobic condition will prevail, where sulphate is used as the electron acceptor. This paper presents the results of a study on anoxic transformation processes in the bulkwater phase of municipal wastewater. Experiments were conducted under conditions of excess electron donor and excess electron acceptor under sewer conditions. The two-stage anoxic transformation process established by Abdul-Talib *et al.* (2001) will serve as an example to establish how the concept of electron equivalent can be used to compare the utilization rate of different types of electron acceptors, i.e., nitrate and nitrite. The study found that the median of electron acceptor utilization rates during Stage I and Stage II are 0.798 e-eq $m^{-3}h^{-1}$ and 0.501e-eq$m^{-3}h^{-1}$ respectively.

Keywords Anoxic condition, electron acceptor, electron donor, electron equivalent, microbial processes

Introduction

Treatment of wastewater during transport in sewer systems is no longer disputed. Laboratory and field evidence on microbial processes during transport in sewer networks have been presented by Green *et al.* (1985), Nielsen *et al.* (1992), Tanaka and Hvitved-Jacobsen (1998), Abdul-Talib *et al.* (2001) and Mourato *et al.* (2002). Microbial transformations occur in sewers when three elements, namely heterotrophic biomass, electron donor and electron acceptor are present. Bentzen *et al.* (1995) established that electron acceptors in wastewater are utilized in a fixed sequence: oxygen for aerobic respiration, nitrate for denitrification, organic compounds for fermentation, sulphate for sulphate reduction and carbon dioxide for methanogenesis. Extensive studies on anoxic transformation processes in municipal wastewater have been conducted by Abdul-Talib *et al.* (2001), Shaharuddin *et al.* (2002) and Rusli and Abdul-Talib (2002). These studies have proven that anoxic transformation of municipal wastewater occurs in two stages as shown in Figure 1.

 Sewer Networks and Processes within Urban Water Systems edited by J.-L. Bertrand-Krajewski, M. Almeida, J. Matos & S. Abdul-Talib. ISBN: 1 84339 506 1.

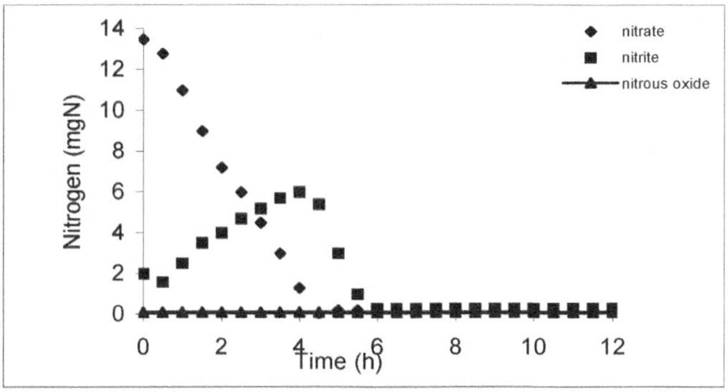

Figure 1 Variations of nitrite and nitrous oxide during anoxic transformation processes in the bulkwater phase (Abdul-Talib *et al.*, 2001)

During Stage I, nitrate is utilized with a significant accumulation of nitrite. Since NO_2-N accumulation peak is less than the initial NO_3-N concentration it can be deduced that NO_2-N is also utilized during Stage I. In Stage II, only NO_2-N remained and served as electron acceptor. Comparison on processes occurring during the two stages cannot be made directly, as the electron acceptors involved are different. Nitrate and nitrite are simultaneously used as electron acceptors during Stage I. However, during Stage II only nitrite is used. Therefore there is a need to establish a method that will enable process rates during the two stages to be compared. This paper introduces the concept of electron equivalent that enables electron utilization rates between the two stages to be compared.

Materials and method

Tests were conducted on 7 different samples taken from the inlet of a municipal wastewater treatment plant (WWTP) located in Section 23, Shah Alam. The WWTP was designed for a population equivalent of 380 000, and receives only domestic wastewater.

Bulkwater samples were taken directly from the inlet of the WWTP so as to ensure that the wastewater has only been subjected to sewer conditions and have not undergone any treatment processes at the WWTP. Besides, approval cannot be obtained for direct sampling from the manholes.

The samples were retrieved using grab sampling method. Samples were filled to the brim in 25-litre containers and tightly seal thus creating an anaerobic condition. Transformation under anaerobic conditions have been reported to be very slow, thus it is reasonable to assume that very little transformation takes place during the 30 minutes transportation from the WWTP to the laboratory(Tanaka and Hvitved-Jacobsen, 1999).

Electron acceptor utilization rate test

Tests were performed to measure electron acceptor utilization rates under conditions of excess electron acceptor and excess electron donor. Excess electron donor conditions were obtained by adding sodium acetate equivalent to 200g COD/m^3 into the sample, while 9-12 g NO_3-N/m^3 were added to create excess electron acceptor conditions. Electron acceptor utilization rates were determined in batch reactors made of clear perspex with a total volume of 2 l. The reactors were filled with 1.5 l of sample, leaving a headspace of 0.5 l as shown in Figure 2.

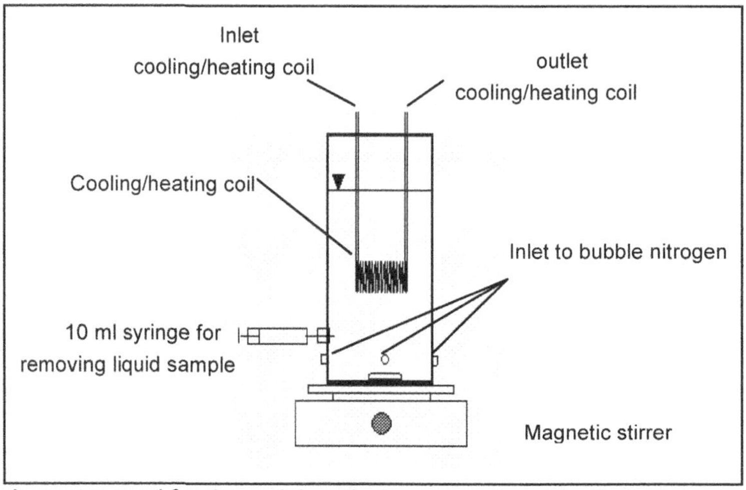

Figure 2 Batch reactor used for determining electron acceptor utilization rate

Nitrogen was bubbled through the sample for at least 3 minutes after the addition of sodium nitrite (electron acceptor) and sodium acetate (electron donor). This is to ensure that the wastewater sample and the headspace in the reactor are free from oxygen that would otherwise inhibit anoxic processes. Magnetic stirrers were used to continuously mix the sample.

A temperature control system comprising of a water bath and water pump that continuously circulate water through copper coils, that are immersed in the reactors is used to ensure that temperature in all reactors is the same as the temperature in the water bath. The temperature in the water bath is kept constant at 27 °C to reflect the sewer conditions in hot climate country.

Sub-samples comprising of 3ml were extracted from the reactor every 20 minutes for the first 2-hours and every 30 minutes after that. Syringes were used to withdraw the sub-samples. The sub-samples were filtered through a 0.7μm glass fiber filter and kept frozen at -20 °C until they were tested for nitrate and nitrite using an ion chromatography unit.

Results and discussion

The variation in concentration of nitrate (NO_3^-) and nitrite (NO_2^-) during anoxic transformation processes can be plotted against time. A typical variation is shown in Figure 3. In order to plot these two different electron acceptors on the same axis, the amount of nitrogen present in these electron acceptors were used, *i.e.*, NO_3-N and NO_2-N.

Results from this study as shown in Figure 3, confirmed the two-stage anoxic transformation processes established by Abdul-Talib *et al.* (2001). Using methods established by Abdul-Talib *et al.* (2001) nitrate utilization rate during two stages can be determined directly from Figure 3. These graphs have been analyzed and the range of nitrate utilization rate during Stage I was found to be between 1.17 to 6.76 g NO3- $Nm^{-3}h^{-1}$. A comparison of nitrate utilization rate during Stage I and that established in literature is given in Table 1.

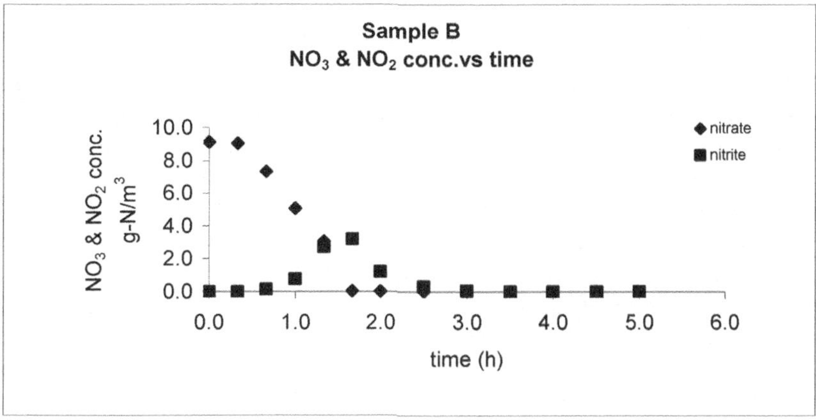

Figure 3 Typical variations of electron acceptors expressed as NO_3-N and NO_2-N

Table 1 Comparison of NUR during stage I in the bulkwater phase established in various studies

Studies	Conditions	NUR e-eq/(m^3h)
Abdul-Talib *et al.* (2002)	Studies on municipal Wastewater in	
	Frejlev Denmark (20°C)	0.9-4.07
	Vest Aalborg (20°C)	2.54-4.62
Rusli and Abdul-Talib (2000)	Studies on Municipal Wastewater in WWTP Section 23, Shah Alam, Malaysia. (27°C)	1.13-2.77
Shaharuddin *et al.* (2002)	Studies on Municipal Wastewater in WWTP Section 1, Shah Alam, Malaysia (27°C)	2.01-3.76
Johari and Abdul-Talib (2004)	Studies on Municipal Wastewater in WWTP Section 1, Shah Alam, Malaysia (27°C)	0.39-1.07
This study	Studies on Municipal Wastewater in WWTP Section 23, Shah Alam, Malaysia (27°C)	1.17-6.76

A close examination of Table 1 shows that the nitrate utilization rate in Malaysian wastewater is highly variable. Even tests conducted on samples taken from the same location yield highly variable results. In general all studies, yielded rates that are in the same order of magnitude. No specific conclusion can be drawn based on these results unless information on the concentration of biomass in all the samples is available.

Microbial processes during the anoxic transformations basically involve redox reactions, where the electron acceptor is reduced and at the same time the electron donor is oxidized.

The reduction in the electron acceptors can be illustrated using the path describing the two-stage anoxic transformation processes as shown below;

$$\underset{(+5)}{NO_3^-} \longrightarrow \underset{(+3)}{NO_2^-} \longrightarrow \underset{(0)}{N_2} \qquad (1)$$

In transforming nitrate to nitrite, the nitrogen atom is reduced from an oxidation state of +5 to +3. Therefore 2 electrons have been transferred to the electron acceptor, nitrate. The transformations of nitrite to nitrogen further reduce the oxidation state of the nitrogen atom from +3 to 0. During this transformation 3 electron were transferred to nitrite. The rate at which electrons were transferred can be used as a measure of the transformation rates during the two-stage anoxic transformation processes. This can be done by manipulating the nitrate and nitrite utilization rates in terms of $gNO_3\text{-}Nm^{-3}h^{-1}$ and $gNO_2\text{-}Nm^{-3}h^{-1}$ to electron equivalents as shown in Figure 4.

The steps shown in Figure 4 were applied to all data in this study and the complete results are given in Table 2. These results allow transformation rates between the two stages, involving different electron acceptors to be compared.

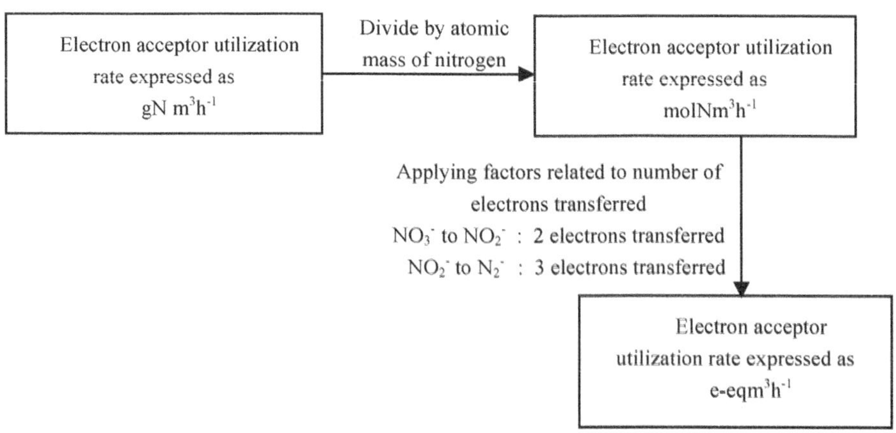

Figure 4 Steps in converting the electron acceptor utilization rates to electron equivalents

Table 2 Electron acceptor utilization rates during anoxic transformation processes under excess donor and excess acceptor conditions

		Stage I		Stage II		Stage I			Stage II
Sample	Nitrate	Nitrite	Nitrite	Nitrite	Nitrate	Nitrite	Nitrite	Nitrate & Nitrite	Nitrite
	UR	AR	UR	UR	UR	AR	UR	UR	UR
		$gNm^{-3}h^{-1}$		$gNm^{-3}h^{-1}$		e-eq$m^{-3}h^{-1}$			e-eq$m^{-3}h^{-1}$
A	4.455	3.811	0.644	5.6	0.636	0.544	0.138	0.774	1.2
B	6.514	2.665	3.849	3.331	0.931	0.381	0.825	1.755	0.714
C	6.755	3.628	3.127	3.48	0.965	0.518	0.67	1.635	0.746
D	4.707	2.259	2.448	2.34	0.672	0.323	0.525	1.197	0.501
E	5.503	4.589	0.914	3.319	0.786	0.656	0.196	0.982	0.711
F	2.457	0.371	2.086	0.931	0.351	0.053	0.447	0.798	0.2
G	1.173	0.293	0.88	1.194	0.168	0.042	0.189	0.356	0.256
H	1.62	0.306	1.314	0.499	0.231	0.044	0.282	0.513	0.107
I	1.592	0.58	1.012	1.106	0.227	0.083	0.217	0.444	0.237
Median	4.455	2.259	1.314	2.34	0.636	0.323	0.282	0.798	0.501
Ave	3.864	2.055	1.808	2.422	0.552	0.294	0.387	0.939	0.519
Std. Dev	2.195	1.716	1.133	1.658	0.313	0.245	0.242	0.504	0.519

Observations made on Table 2 indicated that the utilization rates of electron acceptors were highly variable.

The median and standard deviation of the electron acceptor utilization rates during Stage I expressed in terms of electron equivalent are 0.798e-eq $m^{-3}h^{-1}$ and 0.504e-eq $m^{-3}h^{-1}$ respectively. The corresponding values in Stage II are 0.501 e-eq $m^{-3}h^{-1}$ and 0.355e-eq $m^{-3}h^{-1}$. The variation of electron acceptor utilization rates in all samples during the two stages transformation processes is shown in Figure 6.

Comparison of rates as shown in Figure 6 indicated that the processes in the wastewater samples were highly variable. In all cases, with the exception of sample A, the electron acceptor utilization rate during Stage I is higher than that of Stage II. The ranges of ratio of rates taken between the two stages are 2.46 to 1.39.

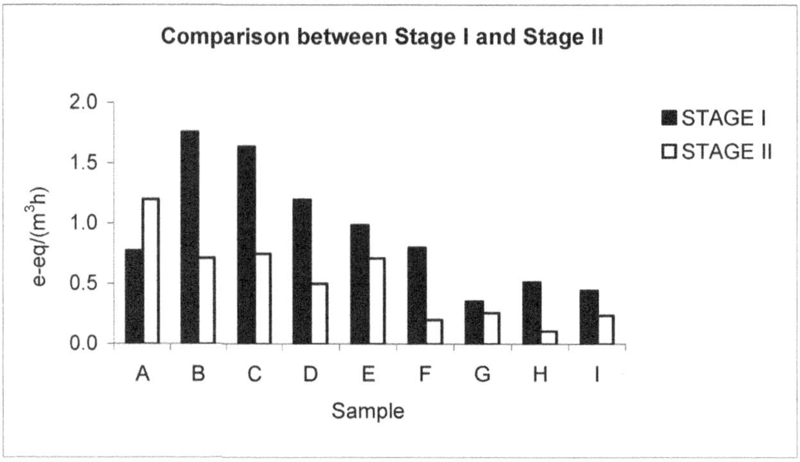

Figure 6 Variation of electron acceptor utilization rates in all samples during the two stage transformation processes

Conclusions

The study on anoxic transformation processes in bulkwater phase in municipal wastewater has confirmed the two-stage process established by Abdul-Talib *et al.* (2001).

In general the electron acceptor utilization rates expressed in terms of e-eq$m^{-3}h^{-1}$ during Stage I is higher than the rates in Stage II. The median of electron acceptor utilization rate during Stage I and Stage II are 0.798 e-eq$m^{-3}h^{-1}$ and 0.501e-eq$m^{-3}h^{-1}$ respectively.

The use of electron equivalent has been successfully used to compare electron utilization rates not only between the different stages but also between processes utilizing different electron acceptors. The concept can be extended to compare processes occurring under different condition, i.e. between aerobic, anoxic and anaerobic conditions where the electron acceptors are dissolved oxygen, nitrate/nitrite and sulphate are the electron acceptors respectively.

References

Abdul-Talib, S., Hvitved-Jacobsen, T., Vollerrtsen, J. and Ujang, Z.(2001). "Anoxic transformations of wastewater organic matter in sewers – process kinetics, model concept and wastewater treatment potential". *Wat. Sci. Tech.* **45**(3), 53-60.

Bentzen, G., Smith, A.T., Bennett, D., Webstar, N.J., Reinholt, F., Sletholt, E. and Hobson, J. (1995). "Controlled dosing of nitrate for prevention of H_2S in a sewer network and the effects on the subsequent treatment process". *Wat. Sci. Tech.* **31**(7), 293-302.

Green, M., Shelef, G. and Messing, A. (1985). "Using the sewerage system main conduits for biological treatment". *Wat. Env. Res.* **19**(8), 1023-1028.

Johari, N.S. and Abdul-Talib, S. (2004). "Nitrate utilization rate in the bulkwater and biofilm phases of sewer network". *1st Junior Scientists Workshop on Sewer Network Processes & Urban Drainage.* 7-10 Feb. Melaka, Malaysia.

Mourato, S., Almeida, M., Matos, J.S. and Hvitved-Jacobsen, T. (2002). "Modelling in-sewer pollutant degradation processes in the costa do Estoril sewer system". *Proc. 3rd. Int. Conf. On Sewer Processes and Networks*, 15-17 June, Paris, France:81-88.

Nielsen, P.H., Raunkjaer, K., Norsker, N.H., Jensen, N.A. and Hvitved-Jacobsen, T. (1992). "Transformation of wastewater in sewer system-a review". *Wat. Sci. Tech.* **25**(6), 17-31.

Rusli, Z. and Abdul-Talib, S. (2002). "In–sewer processes: Microbial transformation under anoxic condition in the bulkwater phase of municipal wastewater". *Proc. of the First Students Congress of Chemical Engineering*, 1st – 2nd February 2002, Universiti Malaysia Sabah, Sabah.

Shaharudin, M., Abdul-Talib, S. and Kamarun, D. (2002). "In-Sewer Processes: Denitrification in the bulkwater phase of municipal wastewater". *Proceeding Seminar Insight 2002*, 24-25 August 2002, PWTC, Kuala Lumpur, 33-41.

Tanaka, N. and Hvitved-Jacobsen, T. (1998). "Transformation of wastewater organic matter in sewers under changing aerobic/anaerobic conditions". *Wat. Sci. Tech.* **37**(1), 105-113.

Tanaka, N. and Hvitved-Jacobsen, T. (1999). Anaerobic transformations of wastewater organic matter under sewer conditions. In: I.B. Joliffe and J.E. Ball (eds.), Proceedings of the 8th International Conferenceon Urban Storm Drainage, Sydney, Australia, August 30-September 3, 1999, 288-296.

Anoxic transformations of wastewater in sewer networks

Weixiao Yang, Jes Vollertsen and Thorkild Hvitved-Jacobsen

Section of Environmental Engineering, Department of Life Sciences, Aalborg University, Sohngaardsholmsvej 57, DK-9000, Aalborg, Denmark

Abstract Anoxic processes can effectively control odour and corrosion in sewer networks. However, the absence of fundamental knowledge on anoxic transformation in sewer limits the engineering application for control measures in sewers. This paper focuses on a basic understanding of anoxic transformations in sewers in terms of process kinetics. A bench scale experimental setup was applied to study the fate of nitrate in both bulk water and biofilm. Experiments conducted show that nitrite builds up. The nitrate reduction in biofilm can be defined as a two-stage reaction. The denitrification rate in the bulk water was determined to 2.1 g N $(m^3\ h)^{-1}$. The specific denitrification surface rate in sewer biofilms was found to be 0.15 g N $(m^2\ h)^{-1}$.

Keywords Biofilm; denitrification; sewer; nitrate; nitrite; odour control

INTRODUCTION

Sewer processes proceed under aerobic, anoxic or anaerobic conditions depending on the type of electron acceptor available. These microbial transformations of organic compounds in wastewater utilize different electron acceptors in a fixed sequence: dissolved oxygen, nitrate/nitrite, organic compounds, sulphate and carbon dioxide (Nielsen et al., 1992). Heterotrophic bacteria use dissolved oxygen to oxidize organic matter in wastewater under aerobic conditions. Similar heterotrophic processes utilize nitrate and nitrite as electron acceptors under anoxic conditions whereas organic matter donates the electrons. Anaerobic processes in sewer include fermentation, sulphide formation and methane generation where organic compounds, sulphate and carbon dioxide are utilized, respectively. Fermentation products and sulphides from sulphate reduction are major odour compounds. Hydrogen sulphide is also a toxic compound and is associated with serious corrosion of concrete and metals. Odour and corrosion problems are frequently found in anaerobic dominated sewers where reaeration and ventilation is insufficient or even absent, i.e. in pressure mains.

Nitrate and nitrite are typically not present in significant concentrations in domestic wastewater and anoxic conditions do not naturally occur. However, denitrification can artificially be introduced by nitrate addition in order to control odour and corrosion problems. The use of nitrate is a well-known methodology for hydrogen sulphide control (Fales, 1929; Hobson and Yang, 2000). The nitrate addition establishes anoxic conditions and therefore suppresses the anaerobic activities of fermentative bacteria and sulphate reducing bacteria. Denitrification is a stepwise reduction of nitrogenous oxides (nitrates and nitrites) to dinitrogen (N_2) with gases nitric oxide (NO), nitrous oxide (N_2O) as intermediates. The most probable pathways of anoxic transformations beginning with nitrate

 Sewer Networks and Processes within Urban Water Systems edited by J.-L. Bertrand-Krajewski, M. Almeida, J. Matos & S. Abdul-Talib. ISBN: 1 84339 506 1.

(NO_3^-) are shown in Eq. (1) (Wild et al., 1994). The numbers in brackets are the oxidation level of the nitrogen atom.

$$\underset{(+5)}{NO_3^-} \rightarrow \underset{(+3)}{NO_2^-} \rightarrow \underset{(+2)}{NO} \rightarrow \underset{(+1)}{N_2O} \rightarrow \underset{(0)}{N_2} \quad (1)$$

Although denitrification has been studied extensively with respect to wastewater treatment plants (WWTPs), limited basic studies have been done on the anoxic transformations under sewer conditions, especially in sewer biofilms. Knowledge on anoxic transformations from activated sludge in treatment systems cannot be transferred indiscriminately since sewer conditions differ significantly from those in the treatment plants. A study by Abdul-Talib (2002) proposes that in-sewer denitrification in the bulk water phase can be simplified into two stages. In the first stage, nitrate is utilized with significant nitrite accumulation. The second stage occurs when nitrate has been depleted and the accumulated nitrite is utilized. Fundamental knowledge on the anoxic kinetics in sewer biofilms is still absent. Consequently, the control of odour and corrosion by nitrate addition is today only empirically estimated. Risk of nitrate overdosing leaves two major problems for the downstream WWTP: an increased nitrogen-load and a shortage of easily degradable organic matter.

Conceptual modeling of anoxic transformations requires improved understanding of relevant process kinetics. This paper provides some of the basic kinetics of denitrification in sewers, addressing both bulk water and biofilm. The results and concepts of this study can be directly applied to optimize nitrate dosing in sewer networks.

MATERIALS AND METHODS

Wastewater sampling

Sewage samples were collected at the inlet of Aalborg East WWTP in Denmark, which has a treatment capacity of 100,000 PE. The respective sewer network extends as much as 30 km. The treatment plant receives both domestic and industrial wastewater, mainly via combined gravity sewers. Pressure mains are, however, also integrated in the sewer network. The wastewater at the sampling site can thus be characterized as sewage of relatively high age that has partly been subjected to anaerobic conditions. Bulk water samples collected were used for batch tests within 30 minutes. The total COD of raw wastewater samples were determined initially according to the APHA *et al.* (1998).

Electron equivalent principle

Microbial transformations in wastewater are by nature electron transfer from organic matter to different electron acceptors. The rates of both aerobic and anoxic processes can thus in terms of units be defined on electron equivalent basis, Figure 1. This electron equivalent principle is applied in the following analysis and results, because it allows a direct comparison of both stoichiometry and kinetics. Figure 1 depicts the electron equivalents for the relevant aerobic and anoxic transformation of wastewater.

Oxygen uptake rate (OUR) determination

Oxygen uptake rates of wastewater were determined in batch reactors as shown in Figure 2. Each batch had a volume of approximately 2 L, continuously mixed by magnetic stirrers and kept at 20 ± 2 °C. Dissolved oxygen (DO) concentrations were measured by a Mettler

Toledo DO probe with Teflon membranes. Compressed air supply switching on and off was controlled by a relay with preset DO levels of 60% and 80% of saturation, respectively. DO concentrations were automatically logged on a computer vs. time. OURs were calculated from these measurements by linear regressions on DO.

	Electron donation	**Electron accepting**	**Unification factor**
Aerobic respiration	$\text{Organic matter} \xrightarrow{-e} CO_2$	$O_2 \xrightarrow{+4e} 2O^{2-}$	**8.0 g O_2 (mol e-eq)$^{-1}$**
Nitrate reduction	$\text{Organic matter} \xrightarrow{-e} CO_2$	$NO_3^- \xrightarrow{+2e} NO_2^-$	**7.0 g NO_3^--N (mol e-eq)$^{-1}$**
Nitrite reduction	$\text{Organic matter} \xrightarrow{-e} CO_2$	$NO_2^- \xrightarrow{+3e} \frac{1}{2} N_2$	**4.7 g NO_2^--N (mol e-eq)$^{-1}$**

*** 1 mole e-eq (mole electron equivalent) is defined as the number of moles electrons exchanged where 1 mole electrons equals Avogadro's number, N_A=6.023 10^{23}**

Figure 1 Electron equivalent transformation factor for aerobic and anoxic microbial processes

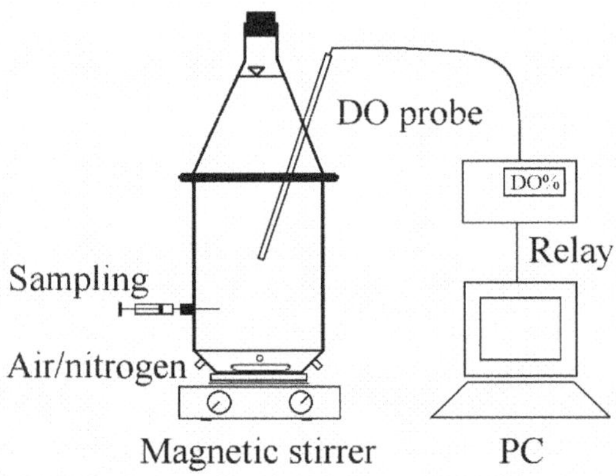

Figure 2 Batch reactor for OUR and denitrification rate determinations in bulk water

According to Hvitved-Jacobsen *et al.* (1998) and Hvitved-Jacobsen (2002), the OUR measurements can be used for identifying the organic matter composition of wastewater (Figure 3). Area A and B are equivalent to the COD fraction of easily degradable and fast hydrolysable organic matter respectively, whereas area C accounts for the slowly hydrolysable organic fraction. It is in this study assumed that easily degradable and fast hydrolysable organic matter serve as the only carbon sources for anoxic transformations (area A+B).

If organic substrate becomes insufficient, degradation of the biomass will supply what is required for the microbial reactions, i.e. the OUR baseline is interpreted as endogenous respiration. It is normally achieved after about 1 day of either aerobic or anoxic conditions of the wastewater.

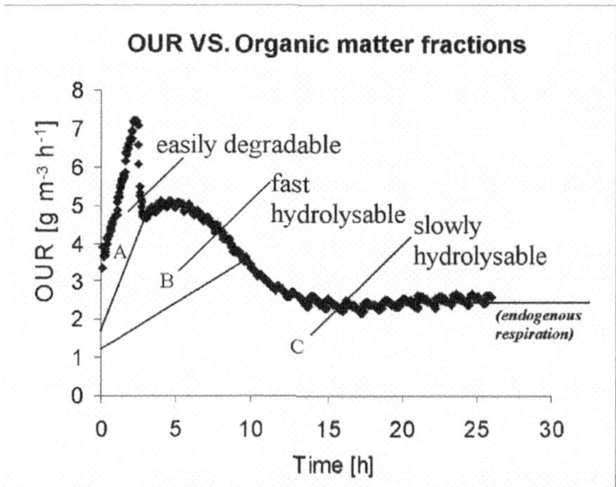

Figure 3 Estimation of organic fractions by OUR simulation (Principle layout)

Nitrate/nitrite utilization rate (NUR) determination

Setup for bulk water studies

Denitrification in bulk water was investigated in a similar way as OUR, Figure 2. Volume, mixing conditions and temperature were also identical with the OUR tests. However, initial DO concentrations were adjusted to a level below 0.5 % of saturation by stripping with nitrogen and consequently no aeration took place during the experiments. Nitrate was added to the system to create anoxic conditions. Liquid samples of 3 ml were taken with equal time intervals during 6 hours and filtered through 0.7 μm glass fiber paper. Nitrate and nitrite were following analyzed by TRAACS 800, a flow injection auto-analyzer based on spectrophotometry principle.

Setup for biofilm phase studies

Three bench scale sewer pipe reactors were constructed to grow sewer biofilms and observe the corresponding anoxic transformations. As shown in Figure 4, two PVC pipes with inner diameter of 21 mm were jointed and connected to a 2-liter reservoir. Wastewater in the reservoir was kept at 20 ± 2 °C and circulated in the pipeline at full flow state. The reservoir was covered with a plastic film to isolate it from possible reaeration during water circulation. By adjusting the inlet valve, steady state hydraulic conditions with specific shear stresses can be achieved in the pipes. In order to observe the growth of biofilm, a 1-meter long pipe was placed as a removable section 3.0 meters from the inlet valve.

Anoxic conditions were initially achieved by stripping nitrogen and adding nitrate. During the preparation of sewer biofilm, fresh wastewater was changed every second day. The total volume of wastewater in both reservoir and the pipeline was substituted with fresh wastewater before each experiment. The sampling and analysis procedures were similar to those of the bulk water phase studies which are described above. The biofilm thickness was estimated by measuring the wet weight within the 1-meter observation pipe. The pipe was drained for 15 minutes and the total weight of pipe and biofilm was measured. Assuming a specific density of 1.0 g cm^{-3}, the biofilm thickness was calculated by subtracting the original weight of the clean pipe. The biofilm growth was only allowed in the pipes. The reservoir was disassembled and all unwanted biofilm growth was removed before wastewater was changed.

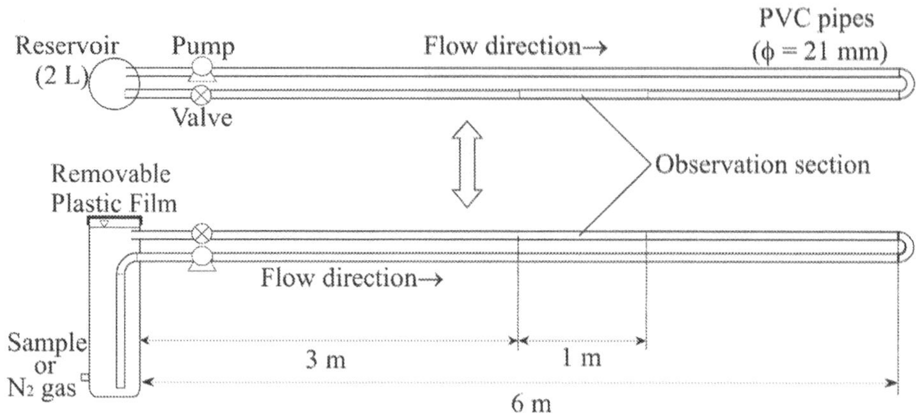

Figure 4 Bench scale setup for observing the anoxic transformation in sewer biofilms

Rate calculation

Nitrate utilization rates are calculated from observed concentrations by linear regressions on NO_3^--N vs. time, Eq. (2). Nitrite utilization rates, however, cannot be calculated simply from nitrite variations. Figure 5 illustrates a typical denitrification scenario taking place in wastewater. When nitrate is present (Stage I), the linear regression on NO_2^--N vs. time reflects only the net accumulation rate of two processes, i.e. nitrite production from nitrate reduction and nitrite utilization, Eq. (3). After nitrate is depleted (Stage II), the nitrite utilization rate is calculated directly from the concentrations by linear regressions on NO_2^--N vs. time, Eq. (4).

Nitrate

$$r_{NO3} = -\frac{dC_{NO3}}{dt} \tag{2}$$

Nitrite utilization rate Stage I:

$$r_{NO2} = r_{NO3} - r'_{NO2} = \frac{dC_{NO2}}{dt} - \frac{dC_{NO3}}{dt} \tag{3}$$

Nitrite utilization rate Stage II:

$$r_{NO2} = r'_{NO2} = -\frac{dC_{NO2}}{dt} \tag{4}$$

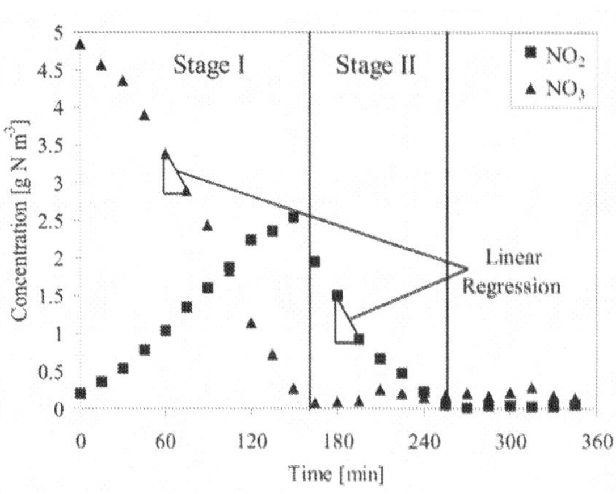

Figure 5 Illustration of nitrite utilization rate determination

In the biofilm reactor, the anoxic transformations take place in both biofilm and bulk water. The nitrate and nitrite utilization rates in the biofilm can therefore be calculated by subtracting the bulk water rates. The NUR in biofilms are often expressed as a surface rate. This specific surface rate is calculated as the surface area of the biofilm by multiplying the rate with an A/V ratio, where V is the volume of bulk water in the reactor and A is the wetted area where biofilm is in contact with the bulk water. In this study, the A/V ratio of a full flowing circular pipe is calculated as Eq. (5).

$$A/V = \frac{2 \cdot \pi \cdot R}{\pi \cdot R^2} = \frac{2}{R} = \frac{4}{D} = 190 m^{-1} \quad (5)$$

Where R and D stands for the radius and diameter of the circular pipe, respectively.

Experimental design

The overall NURs in both biofilm and bulk water were investigated in the biofilm reactors (Fig. 4). Two levels of nitrate addition, resulting in 25 mg NO_3-N l^{-1} and 60 mg NO_3-N l^{-1}, were used. In order to distinguish the anoxic transformations in biofilm, NURs in bulk water were determined in three bulk water reactors (Fig. 2) in parallel with the biofilm reactors. By using identical wastewater and similar sample/analysis procedures, it can be assumed that the NURs of biofilm equal the rate differences between the corresponding biofilm reactor batches and bulk water reactor batches. All the experiment batches used fresh wastewater. OUR batches were conducted as references in order to characterize the basic aerobic kinetics and organic fractions. The experiments conducted are listed in Table 1.

Table 1 Design of the experiment determining NURs in bulk water and biofilm

Waste-water tested	Reactor type	Experiment type	Initial concentration (mg NO_3-N l^{-1})	n*
Batch A	Bulk water reactor (Fig. 2)	Bulk water/NO_3^-	5	3
	Biofilm reactor (Fig. 4)	Biofilm + Bulk water/NO_3^-	25	3
	OUR reactor (Fig. 2)	Bulk water/O_2	-	1
Batch B	Bulk water reactor (Fig. 2)	Bulk water/NO_3^-	5	3
	Biofilm reactor (Fig. 4)	Biofilm + Bulk water/NO_3^-	25	3
	OUR reactor (Fig. 2)	Bulk water/O_2	-	1
Batch C	Bulk water reactor (Fig. 2)	Bulk water/NO_3^-	8	3
	Biofilm reactor (Fig. 4)	Biofilm + Bulk water/NO_3^-	60	3
	OUR reactor (Fig. 2)	Bulk water/O_2	-	1
Batch D	Bulk water reactor (Fig. 2)	Bulk water/NO_3^-	8	3
	Biofilm reactor (Fig. 4)	Biofilm + Bulk water/NO_3^-	60	3
	OUR reactor (Fig. 2)	Bulk water/O_2	-	1

*** n refers to the repeats of the experiment**

RESULTS AND DISCUSSION

A typical example of nitrate utilization in bulk water is shown in Figure 6. The example of concentrations (Fig. 6a) shows that significant nitrite was accumulated within the first 150 minutes and then consumed after nitrate was depleted. This result is in accordance with the two-stage denitrification hypothesis for in-sewer denitrification (Abudul-Talib, 2002). As

NURs are concerned (Fig. 6b), both nitrate and nitrite reduction rates increased within the first 100 minutes while organic substrates and nitrate were abundant. The nitrate reduction rates were higher than the nitrite reduction rates. The nitrate/nitrite reduction rate then started to decrease respectively due to nitrate/nitrite limitation.

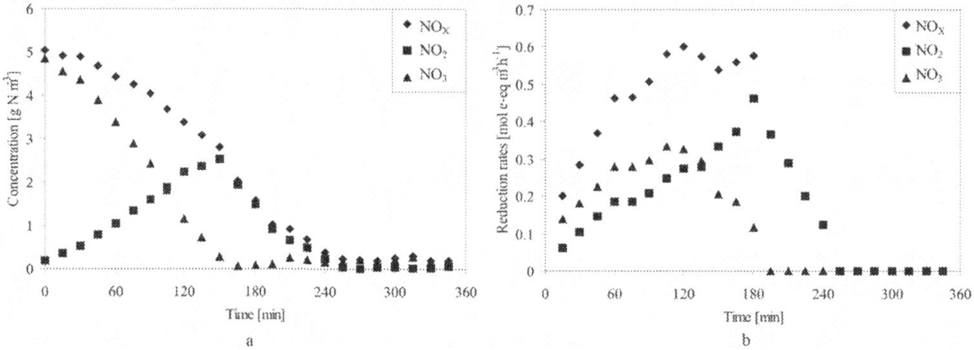

Figure 6 Nitrate utilization of batch A (bulk water)

Figure 7 illustrates typical denitrification scenarios in biofilm reactor batches which describe the overall behavior of both biofilm and bulk water. At both low and high level of initial nitrate concentrations, nitrite accumulations were observed. At low nitrate concentration (Fig. 7a), denitrification can be divided into two stages like in bulk water. Nitrite first built up to a significant level while nitrate is utilized, and was then consumed after nitrate depletion. When the initial nitrate concentration was high (Fig. 7b), both nitrate and nitrite were not depleted within the experimental period. Denitrification ceased after 210 minutes. This is likely due to the lack of readily degradable substrate. The ratio of easily degradable organic substrate to NO_X reduced in this scenario was thus estimated as 9.0 g COD (g N)$^{-1}$. Based on the electron equivalent principle and the definition of yield constant, it can be deduced that the yield for nitrate reduction to nitrogen gas is 0.32 mol e-eq (mol e-eq)$^{-1}$. It also reveals that anoxic transformations rely on the organic substrate content of wastewater. Extended to the engineering point of view, a surplus of nitrate addition will increase the nitrogen load to wastewater treatment and thereby increase the demand of external carbon source for completing denitrification.

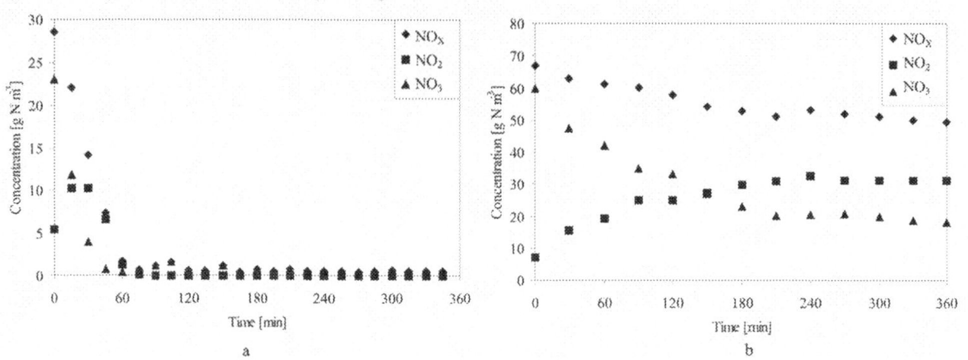

Figure 7 Nitrate utilization of batch B (biofilm + bulk water)

The overall initial anoxic transformation rates in the experiment batches are presented in Table 2. The readily degradable COD differs between wastewaters. The biofilm thickness of every batch can be considered as the same for all the parallel biofilm reactors. The observed nitrate utilization rates are much higher in the biofilm reactors compared with that in the bulk water reactors. The rate differences can be attributed to the anoxic transformations in the biofilm phase. The average anoxic transformation rate in the bulk water is determined to 0.76 mol e-eq $(m^3\ h)^{-1}$, which is equivalent to 2.1 g N $(m^3\ h)^{-1}$ of nitrate reduced to nitrogen gas. The average surface rate of anoxic biofilms under the experimental conditions is calculated to 0.15 g NO_3-N $(m^2\ h)^{-1}$.

Table 2 Basic kinetics of anoxic transformations in both bulk water and biofilm phases

Waste-water tested	Readily degradable COD [mg l^{-1}]	Biofilm thickness [μm]	Transformation rate in bulk water reactors [mol e-eq $(m^3\ h)^{-1}$]	Transformation rate in biofilm reactors [mol e-eq $(m^3\ h)^{-1}$]	Biofilm: calculated surface rate [g N $(m^2\ h)^{-1}$]
Batch A	135	100 ± 5	0.61 ± 0.05	6.2 ± 0.5	0.14
Batch B	205	116 ± 9	0.91 ± 0.08	8.7 ± 0.9	0.18
Batch C	140	134 ± 12	0.65 ± 0.05	6.0 ± 0.6	0.13
Batch D	185	153 ± 13	0.85 ± 0.07	7.9 ± 0.7	0.16
Average			0.76		0.15

As previously described, the sewer biofilm kinetics can be expressed as a surface removal rate of nitrate. Table 3 indicates that the result from this study, 0.15 g NO_3-N $(m^2\ h)^{-1}$, is in good accordance with the literature data, which are obtained from both pilot and full-scale sewer biofilms. The anoxic surface rates in biofilms are much lower than those under aerobic conditions (Norsker *et al.*, 1995).

Table 3 Comparison of specific surface removal rates of sewer biofilms

Reference	Biofilm type	Surface removal rate [g NO_3-N $(m^2\ h)^{-1}$] or [g O_2 $(m^2\ h)^{-1}$]	Remarks
This study	Anoxic/NO_x	0.15	Bench scale sewer biofilm, 20°C
Æsøy *et al.* (1997)	Anoxic/NO_x	0.15-0.19	Pilot scale sewer biofilm, 15°C
Huisman (2001)	Anoxic/NO_x	0.10-0.17	Field study of sewer biofilm, 20°C
Norsker *et al.*, (1995)	Aerobic/O_2	1.2-1.8	Aerobic biofilm feeding with municipal wastewater, 20°C

Assuming the surface rate of biofilm from this study is valid for full flowing circular drainage pipes, the fate of nitrate in sewer can be estimated for different pipes diameters. Figure 8 illustrates the relationship between the sewer nitrate utilization rates and pipe diameters.

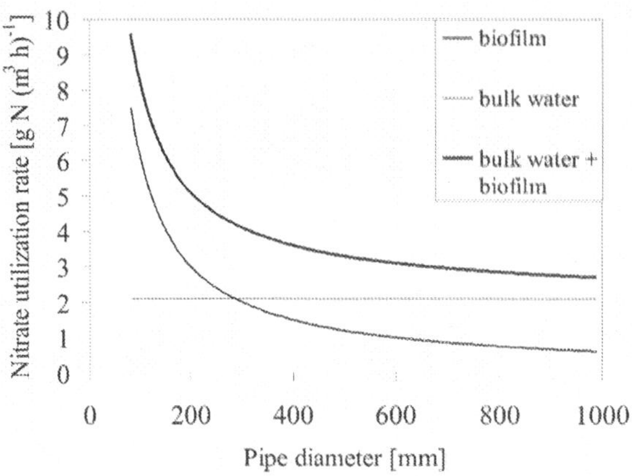

Figure 8 Estimations of nitrate utilizations in sewer networks

CONCLUSIONS

The present study has provided experimental results on basic kinetics of anoxic transformations of raw wastewater under sewer conditions. Both bulk water and biofilm experiments were conducted. The investigations reveal that anoxic process in either bulk water or biofilm must be considered in two steps that correspond to reduction of nitrate and nitrite respectively.

This study also provides an experimental setup that can be used for the study of transformations in biofilms. Under the condition of a high A/V 190 m^{-1}, the anoxic transformations in bulk water can be neglected as compared with that in the biofilm. Based on the comparison of the surface rates, it may be concluded that nitrate reduction in the anoxic biofilm is much slower than oxygen consumptions in the aerobic biofilm.

REFERENCES

Abdul-Talib, S. (2002). "In Sewer Processes: Transformation of Organic Matter under Anoxic Conditions", University Technology Malaysia, Ph.D. Thesis.

APHA, AWWA, WEF (1998). *Standard Methods for the Examination of Water and Wastewater* 20th edn, Washington DC, USA.

Fales, A.L. (1929). Treatment of industrial waste from paper mills and tannery on Neponsit River. Jour. *Ind. Eng. Chem.*, **21**, 216.

Hobson, J. and Yang, G. (2000). Odour control and practical experiences - The ability of selected chemicals to suppress odour formation in a rising main. *Wat. Sci. Tech.*, **41**(6), 165-174.

Huisman, J.L. (2001). Transport and transformation processes in combined sewers. Ph. D thesis, ETH Zurich, ZHW Schriftenreihe Band 10, Zurich, Switzerland.

Hvitved-Jacobsen, T., Vollertsen, J. and Nielsen P.H. (1998). A process and model concept for microbial wastewater transformations in gravity sewers. *Wat. Sci. Tech.*, **37**(1), 233-241.

Hvitved-Jacobsen, T. (2002). *Sewer Processes: Microbial and Chemical Process Engineering of Sewer Networks*, CRC press.

Nielsen, P.H., K Raunkjaer, N.H.Norsker, N.Aa. Jansen and T. Hvitved-Jacobsen (1992). Transformation of wastwater in sewer system-a reveiw. *Wat. Sci. Tech.*, **25**(6), 17-31.

Norsker N. H., Nielsen P. H. and Hvitved-Jacobsen T. (1995). Influence of oxygen on biofilm growth and potential sulfate reduction in gravity sewer biofilm. *Wat. Sci. Tech.*, **31**, 159-167.

Vollertsen, J. and Hvitved-Jacobsen, T. (1999). Stoichiometric and kinetic model parameters for microbial transformations of suspended solids in combined sewer systems. *Wat. Res.*, **33**(14), 3127-3141.

Wild, D, Schulthess, R von; Gujer, W (1994). Synthesis of denitrification enzymes in activated sludge: Modelling with structured biomass. *Wat. Sci. Tech.*, **30**(6), 113-122.

Nitrate Utilization Rate in the Bulkwater and Biofilms Phases of Sewer Networks

Noora Samsina Johari and Suhaimi Abdul-Talib

Faculty of Civil Engineering, Universiti Teknologi MARA, 40450 Shah Alam, Selangor, Malaysia

Abstract Significant progress has been made in developing fundamental knowledge on kinetics of microbial transformations during transport of wastewater in sewer networks. A model concept has been established for transformations occurring in sewer networks under aerobic and anaerobic conditions (Vollertsen *et al.*, 2001). Studies on in-sewer transformation processes under anoxic conditions have been initiated by Abdul-Talib *et al.* (2002). To date extensive investigations on kinetics of anoxic transformation processes have been conducted on the bulkwater phase. However, very limited information is currently available on kinetics of anoxic transformation in the biofilm phase. This study establishes a methodology for determining nitrate utilization rate (NUR) in bulkwater and biofilm phases. Tests were conducted using batch reactors with biofilm grown under sewer conditions. Raw municipal wastewater samples, taken directly from a sewer pipe or from the inlet of wastewater treatment plant (WWTP), were used in this series of experiments. Nitrate utilization rate (NUR) in bulkwater phase was observed to be in the range of 0.39 to 1.07 $gNO_3\text{-}N\ m^{-3}h^{-1}$, while NUR in the biofilm phase was found to be 0.04 to 0.10 $gNO_3\text{-}N\ m^{-2}h^{-1}$.

Keywords Anoxic conditions, nitrate utilization rate, sewer biofilm

Introduction

The sewerage system has conventionally been considered to comprise basically of sewage and wastewater sources, sewer networks, a wastewater treatment plant and the receiving environment. The sewer network is designed to collect sewage and wastewater from various sources and transport them to a wastewater treatment plant. The wastewater treatment plant is designed to remove/reduce pollutants to an acceptable level before discharging to the receiving environment. The conventional approach which, account for the mass balance of the flow (hydraulic continuity) between the different parts of the sewerage system is well recognized as far as the hydraulics and transport phenomena are concerned. However, the concept has seldom been applied when considering the biological processes.

In order to move towards a more integrated approach for urban wastewater management, the operation and design of sewer networks and wastewater treatment plants must be integrated, incorporating not only hydraulic continuity but also processes continuity. This means that microbial changes occurring during transport in sewer networks must be accounted for and integrated to the processes that occur at the receiving wastewater treatment plant. Intense studies have been directed towards understanding microbial transformation processes during transport in sewers. Evidence from field investigations on microbial transformation in sewers leading to wastewater quality changes have been reported over the last fifteen years (Green *et al.*, 1985; Almeida, 1999; Huisman, 2001). Transformations of

 Sewer Networks and Processes within Urban Water Systems edited by J.-L. Bertrand-Krajewski, M. Almeida, J. Matos & S. Abdul-Talib. ISBN: 1 84339 506 1.

organic matter in sewers involving mass transfer occur simultaneously between the different phases of the sewer environment, namely, the bulkwater, the sewer biofilm, the sewer sediment and the sewer atmosphere phases. These transformations may occur either under aerobic, anoxic or anaerobic conditions depending on the availability of the type of electron acceptors (Nielsen *et al.*, 1992). The electron acceptors are utilized in a fixed sequence: oxygen for aerobic respiration, nitrate for denitrification, organic compounds for fermentation, sulphate for sulphate reduction and carbon dioxide for methanogenesis (Bentzen *et al.*, 1995).

Studies to establish fundamental knowledge on processes occurring in the different phases of the sewer environment have been conducted by several researchers. Bjerre (1996) and Vollertsen (1998) established model concepts for processes occurring under aerobic conditions in the bulkwater and sewer sediment phases. Investigation into processes under changing anaerobic/aerobic conditions in the bulkwater phase was reported by Tanaka (1998), while fundamental concepts governing anoxic transformation processes in the bulkwater phase have been reported by Abdul-Talib *et al.* (2002).

A search in literature has indicated that very few studies have been directed towards establishing the fundamental knowledge on transformations in the biofilm phase – especially on anoxic transformations. The objective of this paper is to present a methodology for studying the kinetics of processes in sewer biofilm under anoxic conditions. Nitrate utilization rates in the bulkwater and biofilm phases will be established and compared.

Materials and Method

Sampling Location

Raw municipal wastewater samples were used during this study. Samples were collected directly from the inlet of a wastewater treatment plant (WWTP) located at Jalan Ilmu, Universiti Teknologi MARA, Shah Alam. Grab sampling method was used to retrieved the samples. The samples were taken into analysis within 20-30 minutes of sampling and were kept anaerobic to minimize microbial transformation processes during transport to the laboratory

Bulkwater and Biofilm Reactor

A series of laboratory tests were conducted using batch reactors specially designed for this study. The two types of reactors used in this study – designated as Type A and Type B are schematically shown in Figures 1 and 2. Both reactors were made from PVC pipe material. Appropriate fittings such as sampling port, cooling/heating coil and the inlet of nitrogen gas were incorporated to enable nitrate and nitrate utilization rates in both reactors to be monitored.

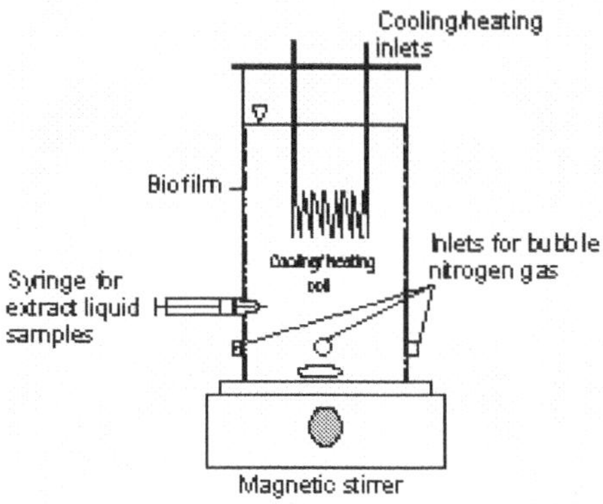

Figure 1 Type-A reactor (with biofilm)

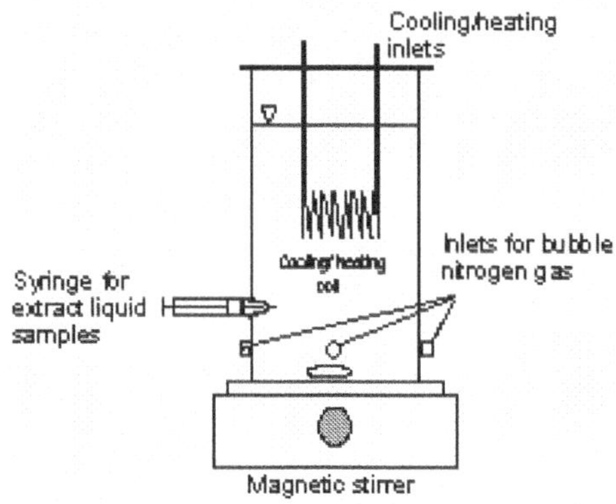

Figure 2 Type-B reactor (free of biofilm)

Preparation of sewer biofilm

The biofilm was grown in Type-A reactor under sewer conditions. Raw municipal wastewater was fed into the reactor, subjected to alternating anoxic and anaerobic conditions. Bulkwater was replaced every two day and supplemented with sodium nitrate and sodium acetate to acclimatize the biofilm to anoxic conditions. A duration of two months was allowed for the biofilm to be fully developed.

Laboratory experiments

In total, experiments were conducted on 6 different wastewater samples. Both reactors were filled with 6 litres of bulkwater samples. Nitrogen gas was bubbled into the reactor to remove dissolved oxygen in the bulkwater sample and in the headspace of the reactor. Anoxic conditions were created by the addition of sodium nitrate. Excess substrate conditions were achieved through the addition of sodium acetate. Liquid samples were extracted from both reactors using a syringe. The samples were extracted every 20 minutes for the first 2 hour and every 30 minutes for the remaining of the test duration. The extracted samples were immediately filtered using 0.7μm glass fibre filter. The samples were kept frozen at -20°C until they were tested for nitrate and nitrite. Magnetic stirrers were used during the experiment to ensure complete mixing of the samples throughout the experiments. The temperature in both reactors was kept constant by water from a water bath that flowed continuously through the cooling coils fitted to both reactors.

Measurement of nitrate and nitrite

Nitrate and nitrite concentrations were determined using Ion Chromatography unit (790 IC). Ion Chromatography unit is equipped with 5 μm particle size column packing of polyvinyl alcohol with quaternary ammonium groups and eluent used as a carrier. The eluent consist of 3.2m mol/L Na_2CO_3 and 1.0 m mol/L $NaHCO_3$. The flow rate adopted is 0.7 m L/min

Results and Discussion

Typical variations of nitrate and nitrite concentrations during the anoxic transformation processes in Type A and Type B reactors are shown in Figure 3. A two-stage denitrification process was observed in both reactors. During stage I, nitrate is utilized with a significant accumulation of nitrite. Nitrite accumulation was observed to peak when nitrate is depleted. During stage II, the accumulated nitrite was utilized. The two-stage denitrification process in the bulkwater phase agrees with studies conducted on Danish municipal wastewater reported by Abdul-Talib *et al.* (2002) and studies on Malaysian municipal wastewater reported by Shaharuddin and Abdul-Talib (2002).

Nitrate utilization rates (NUR) in both reactors can be determined directly from Figure 3. In Type A reactor, where both bulkwater and biofilm are present, the NUR were found to be higher than the rates in Type B reactor, where only the bulkwater is present. This is expected as Type A reactor has active biomass from both the bulkwater and biofilm phases compared to Type B reactor where source of active biomass comes only from the bulkwater phase.

As the bulkwater sources were the same for both reactors, it is reasonable to assume that both reactors have the same amount of biomass in the bulkwater phase. Thus the higher NUR observed in Type A reactor must be attributed to the additional active biomass from the biofilm phase. The difference between the NUR in the two reactors gives the NUR in the biofilm phase.

It should be noted that the NUR in the biofilm phase derived in this manner is expressed in terms of $gNO_3\text{-}Nm^{-3}h^{-1}$. The rates can be multiplied by the A/V ratio to convert the deduced NUR in the biofilm phase into g $NO_3\text{-}Nm^{-2}h^{-1}$, where A is the surface area covered by biofilm and V is the volume of the bulkwater.

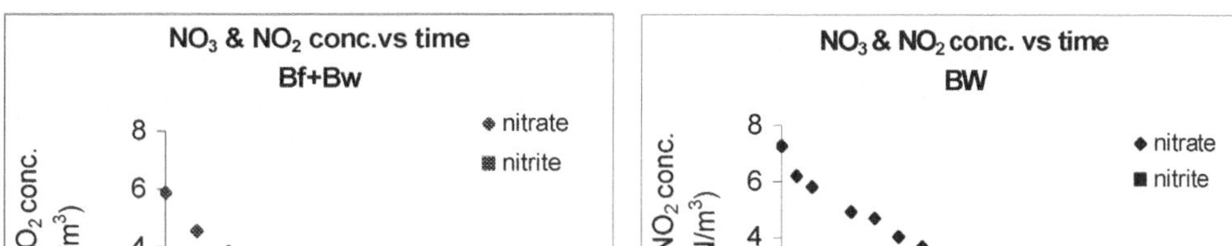

Figure 3 Typical results on nitrate and nitrite concentrations in biofilm and bulkwater and bulkwater only.

A summary of NUR for both reactors and the deduced NUR in the biofilm phase is given in Table 1. A closer look at Table 1 reveals that the NUR in the bulkwater phase ranges from 0.39 to 1.07 $gNO_3\text{-}Nm^{-3}h^{-1}$. Furthermore, with the exception of the first sample, the NUR in the biofilm phase is much higher compared to the NUR in the bulkwater phase. This indicates either there is more active biomass in the biofilm phase or the biomass in the biofilm phase is better acclimatized to anoxic conditions. The NUR in the biofilm phase established in this study ranges from 0.04 to 0.1 $gNO_3\text{-}Nm^{-2}h^{-1}$.

Table 1 Summary of results on nitrate utilization rates (NUR)

Sample	NUR [g NO_3 -$Nm^{-3}h^{-1}$] Reactor A (R_A) *(Bw+Bf)*	Reactor B (R_B) *(Bw)*	R_A-R_B *Bf*	NUR [g NO_3 -$Nm^{-2}h^{-1}$] * *Biofilm*
05.02.02	1.75	1.02	0.73	0.04
08.02.03	2.66	0.87	1.79	0.09
15.02.03	2.79	1.07	1.72	0.09
18.02.03	2.96	1.00	1.96	0.10
26.02.03	1.76	0.39	1.37	0.07
27.02.03	2.45	1.02	1.43	0.07
Median	2.46	1.02	1.44	0.07
Average	2.40	0.90	1.50	0.08
Std. Deviation	0.51	0.41	0.44	0.02

*The NUR in biofilm phase can be expressed in term of gNO_3-$Nm^{-2}h^{-1}$ by multiplying with A/V ratio where A is area where biofilm was grown and V is the volume of bulkwater in the reactor.

Table 2 gives a comparison of the NUR in the bulkwater established in this study against that reported in literature. The NUR in the bulk water phase observed in this study was found to be much lower when compared to other studies. It is possible that biomass in the biofilm phase which are acclimatized to anoxic conditions prior to the experiments, out-perform the biomass in the bulkwater phase, which may require some time to acclimatize to the anoxic conditions.

A similar comparison as given in Table 3 is made on the NUR in the biofilm phase. No direct comparison can be made on the NUR in the biofilm phase reported in studies shown in Table 3. However a general observation indicates that the NUR in biofilm established in studies conducted in an actual sewer pipe tend to be in the range of 0.1-0.2 gNO_3-$Nm^{-2}h^{-1}$. Studies on NUR in batch reactors yield rates that are less than 0.1 gNO_3-$Nm^{-2}h^{-1}$. This can be explained by the fact that biofilm in sewer pipes are more diverse such that there may be more species that are capable of utilizing nitrate. The limited study conducted thus far has provided an insight into microbial processes in the biofilm phase.

Table 2 Comparison of NUR in the bulkwater phase

Studies	Conditions	NUR (g NO_3-N $m^{-3}h^{-1}$)
Abdul-Talib *et al.* (2002)	Studies on Municipal Wastewater in	
	Frejlev Denmark	0.9 – 4.07
	Vest Aalborg	2.54 – 4.62
Rusli and Abdul-Talib (2002)	Studies on Municipal Wastewater in Malaysia	1.14 – 2.77
Shaharuddin *et al.* (2002)	Studies on Municipal Wastewater in Malaysia	2.01 – 3.76
This Study	Studies on Municipal Wastewater in Malaysia	0.39-1.07

Table 3 Comparison of NUR in biofilms

Studies	Conditions	NUR ($g\ NO_3\text{-}N\ m^{-2}h^{-1}$)
Jansen and Harremoes(1984)	Fixed film at WWTP	0.15
Manadhar and Schroder (1995)	NUR in reactors using synthetic wastewater at 30 °C	0.02 – 0.03
Æsøy *et al.*, (1997)	Pilot study in anoxic sewer at 15 °C	0.15 – 0.19
Huisman (2001)	Field studies in sewer biofilm at 20 °C	
	-thin biofilm	0.17
	-thick biofilm	0.1
Poulsen (1997) in Hvitved-Jacobsen (2002)	Biofilm originating from a biofilter at 20 °C	0.025 – 0.055
Abdul-Talib (2002)	Biofilm grown in batch reactors under sewer condition at 27 °C	0.05 – 0.13
This Study	Biofilm grown in batch reactors under sewer condition at 27 °C	0.04 – 0.10

Conclusions

The methodology adopted in this study has enabled NUR to be determined in both the bulkwater and biofilm phases. The NUR in the bulkwater phase established in this study were found to be in the range of 0.39 – 1.07 $gNO_3\text{-}Nm^{-3}h^{-1}$, while the corresponding range in the biofilm phase was found to be between 0.04 and 0.1 $gNO_3\text{-}Nm^{-2}h^{-1}$. Further studies need to be designed to enable a direct determination of NUR in the biofilm phase.

References

Abdul-Talib, S. (2002). In Sewer Processes: Transformation of organic matter under anoxic conditions. Faculty of Civil Engineering, Universiti Teknologi Malaysia: Ph.D. Thesis

Abdul-Talib, S., Hvitved-Jacobsen, T., Vollertsen, J. and Ujang, Z. (2002). Anoxic transformation of wastewater organic matter in sewers-process kinetics, model concept and wastewater treatment potential. *Wat. Sci. Tech.* **45**(3), 53-60.

Almeida, M. (1999). Pollutant transformation processes in sewers under aerobic dry weather flow conditions. Department of Civil and Environmental Engineering, Imperial College of Science, Technology and Medicine, UK:Ph.D. Thesis.

Bentzen, G. Smith, A.T., Bennett, D., Webster, N.J., Reinholt, R., Sletholt, E. and Hobson, J. (1995). Controlled dosing of nitrate for prevention of H_2S in a sewer network and the effect of the subsequent treatment processes. *Wat. Sci. Tech.* **31**(7), 293-302.

Bjerre, H.L. (1996). Transformations of wastewater in an open sewer : The Emscher River, Germany. Environmental Engineering Laboratory, Aalborg University: Ph.D. Dissertation.

Green, M., Shelef, G. and Messing, A. (1985). Using the sewerage system main conduits for biological treatment". *Wat. Res.* **19**(8), 1023-1028.

Huisman, J.L. (2001). Transport and transformation processes in combined sewers. Swiss Federal Institute of Technology, Zurich. Switzerland: Ph.D. Thesis.

Jansen, L.C.J. and Harremoes, P. (1984). Removal of soluble substrates in fixed films. *Wat. Sci. Tech.* **17**(2-8) part 1. 1-14.

Manadhar, U.K. and Schroder, H. (1995). Sewage circulating reactor: An approach to recirculating wastewater in sewers. *Environ. Technol.* **16**(3), 201-211.

Nielsen, P.H., Raunkjaer. K., Norsker, N.H., Jensen, N.A. and Hvitved-Jacobsen. T. (1992). Transformation of wastewater in sewer system – a review. *Wat. Sci. Tech.* **25**(6), 17-31.

Poulsen, B.K. (1997). Anoxisk omsaetning af organisk stof in biofiltre (Anoxic transformations of organic matter in biofilters). Department of Environmental Technology, Technical University of Denmark. M.Sc. Thesis (in Danish), p.56. in Hvitved-Jacobsen, T. (2002). Sewer processes-Microbial and chemical process engineering of sewer networks." Boca Raton, Florida: CRC Press. 237 pp.

Rusli, Z. and Abdul-Talib, S. (2002). In-Sewer Processes: Microbial transformation under anoxic condition in the bulkwater phase of municipal wastewater. *First Students Chemical Engineering Congress (Chem-EC 2002) Programme*, February 1-2, Universiti Malaysia Sabah (UMS), Sabah.

Shaharuddin, M., Abdul-Talib, S. and Kamarun, D. (2002). In-sewer processes: Denitrification in the bulkwater phase of municipal wastewater. *Prosiding Seminar Insight*, 24-25 August 2002, PWTC, Kuala Lumpur, 34-41

Tanaka, N. (1998). Aerobic/anaerobic process transition and interactions in sewers. Environmental Engineering Laboratory, Aalborg University, Denmark: Ph.D. Dissertation.

Vollertsen, J. (1998). Solids in combined sewers- characterization and transformation. Environmental Engineering Laboratory, Aalborg University, Denmark: Ph.D. Dissertation.

Vollertsen, J., Hvitved-Jacobsen, T., Ujang, Z. and Abdul-Talib, S. (2001). Integrated design of sewers and wastewater treatment plants. *Proc. of. IWA Conf. Water and Wastewater Management for Developing Countries*. Kuala Lumpur, Malaysia: 29-31 October, 186-193.

Æsøy, A., Strorfjell, M., Mellgren, L., Helness, H., Thorvaldsen, G., Ødegaard, H. and Bentzen, G. (1997). A comparison of biofilm growth and water quality changes in sewers with anoxic and anaerobic (septic) conditions. *Wat. Sci. Tech.* **36**(1), 303-310.

Infiltration and exfiltration from sewer network with a focus on house connections in city of Prague

I. Princ * and D. Kohout*

* Czech Technical University in Prague, Faculty of Civil Engineering, Laboratory of Ecological Risks in Urban Drainage - LERMO, Thakurova 7, 166 29 Prague 6, E-mails *princ@lermo.cz*, *kohoutd@lermo.cz*)

Abstract Sewer systems constitute a very significant patrimony in European cities. Their structural quality and functional efficiency are key parameters to guarantee the transfer of domestic and industrial wastewater to treatment plants without infiltration or exfiltration. House sewer connection (hereinafter referred to as HC) is standpipe which draining the house estate to the main sewer. In case of combined sewer system is one house connection per one house estate. House connections are the most sensitive parts of whole sewer system. Frequently are neglected during their construction and operational conditions. The paper shows analysis of several damages occurring on the house connections, their causes and influence on the amount of infiltration water to sewer system by HC or exfiltration of wastewater from HC to environment. Thereinafter, the paper focuses on a field experiments and evaluation of results. The measurements of the amount of water infiltrated to a HC or exfiltrated from HC comprises the identification of the defect and the measurement itself. Related Infiltration assessment, the methodology is based on the monitoring of minimal discharges during the night in a catchment with using of Parshall´s flume. In the case of Exfiltration, the methodology is based on a watertightness test. Measurements are performed on a selected catchments in Prague city, in order to obtain a lot of results from different types of build-up area, age of pipes, geology and hydrology conditions. However, in this paper is presented two localities for Infiltration and Exfiltration assessment.

Keywords damages, exfiltration, house connections, infiltration, sewer system

Introduction

The topics of infiltration and exfiltration to and from house connections are included in the project entitled "Assessing Infiltration and Exfiltration on the Performance of an Urban Sewer System – APUSS". This project forms a part of the Fifth Framework Programme of the European Union, and is divided into three work areas covered by 12 work teams. The project has brought together 10 partners in 7 European countries.

House connections are often the most neglected components of the whole sewer system. Not only results of a research in the USA [WEF, 1999] show that they may contribute to as much as 50% of total infiltration of groundwater, even a practical experiences of the Sewer Net Administrators.

Only water with a degree of pollution corresponding to wastewater, and which can be processed by a wastewater treatment plant, may enter the sewer system. The sewer system must be protected from ingress of ballast water.

 Sewer Networks and Processes within Urban Water Systems edited by J.-L. Bertrand-Krajewski, M. Almeida, J. Matos & S. Abdul-Talib. ISBN: 1 84339 506 1.

Exfiltration from leaky house connections can be substantial and of grave importance in terms of protecting groundwater against pollution as well as preserving foundations of buildings. The general constructional state of HC pipe, including point of connection to the main sewer, corresponds to potential leakage amount of waste water to the soil environment.

The Infiltration issue

Ballast water comprises all groundwater entering the sewer system through infiltration, i.e. seeping from the surroundings (mostly subsoil) into damaged sewers or house connections via crevices, apertures or leaking joints in places below the level of groundwater or near substantial escapes from water supply pipes; these are classified as diffuse sources of infiltration. Another type of infiltration is from local limited sources such as brook mouths, drainage systems, water reservoirs or fountain overflows, or cooling water. Ballast water can be classified by time-dependent occurrence – by regular and irregular flows.

Rehabilitation work aimed at achieving a decrease in water infiltrated from surrounding subsoil concentrates mainly on the core sewers thus leaving the house extensions neglected in the majority of cases. Yet, in consideration of the share of house extensions in the total length of the sewer system (21 per cent of the length of the sewer system in Prague), there comes the question of the actual contribution of the house extensions to the volume of infiltration of ballast water. Curtis and Krutsch (1993), for example, in their study for the Sewerage Agency of Washington County, stated that their studied house extensions had as much as a 50 per cent share in infiltration of ballast water.

Conditions of Infiltration into House Connections:

(1) The level of groundwater is above or at least at the level of the house extension
(2) Subsoil around the extension is sufficiently permeable
(3) Existence of a point of infiltration – damage of sewer piping (leaky joints, crevices, etc.)

Infiltration of ballast water increases the demand on the capacity of sewers and water treatment plants. Increased inflow into wastewater treatment plants, thus overloading the plants, something which caused deterioration of function of the plants. Low temperature and dissolution of wastewater have a negative impact on the efficiency of the biological cleansing processes.

The Exfiltration issue

The main problem of the exfiltration issue is contamination of groundwater and springs as sources of water, including drinking water. Wastewater, which leaking from any type of damage pipe of house connections, run through the soil environment and directly threat human health.

From the constructional point of view, water flow in the soil environment continuously washes separate fractions of the soil, creating caverns threatening the stability of building foundations, municipal engineering networks and communications.

Outflow of wastewater from sewers causes an decrease of solid particles transport velocity and resulting faster pipe silting and increasing costs of sewer system maintenance.

Exfiltration Theory

The guiding formula for exfiltration is expressed as a function of the hole outlet flow and flow through the soil environment.

The hole outlet for unsteady flow is $Q = f(t)$, Q and v is not constant:

$$v = \varphi \cdot \sqrt{2gh} \Rightarrow Q = \varphi \cdot \varepsilon \cdot S \cdot \sqrt{2gh}, \tag{1}$$

where φ is a velocity coefficient, ε is a coefficient of contraction, S is the cross-section area (m^2), Q is hole outlet flow ($m^3.s^{-1}$).

Water flow in a variably saturated porous soil environment is defined (by Richards):

$$\frac{\partial \theta}{\partial t} = div(K \cdot gradh + K_0), \tag{2}$$

$$\frac{\partial \theta}{\partial t} = \frac{\partial}{\partial z}\left(K \cdot \frac{\partial h}{\partial z} + K\right), \tag{3}$$

where θ is soil humidity, K is hydraulic conductivity (m/d), h is head pressure (m), z is vertical z-axis.

The procedure mentioned above gives the solution of the total exfiltration amount in equations (1), (2) or (3). This possibility of the mathematical derivation to be applied in case of sufficient inputs to equations. However, the procedure must be verified by measurements on a selected experimental catchment.

Limiting conditions (maximum exfiltration):

(1) Total outflow from house: from all wastewater sanitary devices plus rainwater off the roof plus any illegal connections of drainage water
(2) Backflow of wastewater from the main sewer into the house connections (reflux stopper are recommended only)

Conditions of exfiltration from house extensions:

(1) Groundwater table is below the house connections piping
(2) Damaged piping
(3) The house connections are installed on at least partially permeable backfill and in an unsaturated environment. Exfiltration does not occur with piping installed in impermeable backfill, such as clay. Exfiltration also does not occur in a completely saturated environment, e.g. after heavy rain.

Objective

The objective is to assess and quantify the volumes of groundwater entering to the house connections (infiltration) and the volumes of wastewater leaking from house connections (exfiltration). Quantification of infiltration and exfiltration is based on field measurements on experimental sites as well as on a physical laboratory model. Another aim of the project is to develop new measurement methods to be tested on local catchment and verified on larger catchment of several cities.

Another outcome of the whole APUSS project is to develop decision-making support for administrators and operators to assess the performance of their sewer systems and to choose investment strategies for rehabilitation and repair.

Types of defects and their causes

First, a list of the most frequent defects and their causes was made, as they occur on house connections. A relation matrix was developed indicating insignificant and most significant defects that require attention (Figure 1.1). Mutual interactions are classified using a five-grade scale (with 1 being the emergency condition and 5 describing a small defect) and are used for specification of sequences of individual causes. The scale is similar to the assessment resulting from defects discovered during the camera-aided inspection (CCTV) in accordance with the Standard ATV M143/2a.

Types of defects are the following: leaking joints (at socket, at manhole, at connection of the main sewer), a broken and loose pipe wall, obstacles (offset piping, tree roots, sediments, incrustation), cracks, corrosion, deformations. Causal factors are: level of groundwater in relation to piping, permeability of soil environment, type of material, age, traffic load, service life span, implementation, rainy or dry season aspect.

legend :	
1	100%
2	75%
3	50%
4	25%
5	0%

	damages														factors	
	leakage joints			blocking				corrosion		cracks			others			
factors	bad joints	connections m. sewer	connections manhole	sediments	incrustation	roots	overlap pipe to m sew.	outside	inside	lengthwise	crosswise	radial	abrasion	deformation	**average**	**order of importance**
high permeability	1	1	1	4	5	4	5	3	5	1	1	1	5	3	2,9	5
med permeability	3	3	3	5	5	4	5	4	5	3	3	3	5	5	4,0	13
material	2	2	2	5	5	4	5	5	5	3	3	5	5	2	3,8	12
area	3	3	3	3	4	1	5	4	4	4	4	4	4	3	3,5	8
rain	1	1	1	1	1	3	1	5	5	3	3	4	3	5	2,6	3
life time	1	1	1	3	1	1	4	3	3	1	1	1	3	1	1,8	1
maintainance	3	3	3	3	3	4	3	3	3	1	1	1	3	5	2,8	4
implemantation	1	1	1	5	5	-	3	5	-	4	2	5	5	1	3,2	6
aggressive backgroun	3	3	3	5	5	5	5	1	-	3	3	3	5	5	3,8	11
aggressive water	4	4	4	4	1	5	5	5	1	3	3	3	3	5	3,6	10
heavy traffic	1	1	1	5	5	5	4	5	5	1	1	5	5	1	3,2	7
pulsating GWL	3	3	3	5	5	3	3	3	5	4	1	5	5	1	3,5	8
ege	2	2	2	4	2	3	5	2	2	2	2	2	2	2	2,4	2
damages: **average**	2,2	2,2	2,2	4,0	3,6	3,5	4,1	3,7	3,9	2,5	2,2	3,2	4,1	3,0		
damages: **order of importance**	1	1	1	12	9	8	13	10	11	5	1	7	13	6		

Figure 1.1 Detailed matrix – example for exfiltration.

The most significant defect is at the joining of the house connections and the main sewer. That is caused by poor quality of installation of the piping, causing exfiltration. The subsoil backfill is insufficiently compacted with vertical loading (e.g. road traffic) resulting in compression of both pipes creating a gap. The use of stoneware pipes is advantageous due to higher resistance compared to PVC pipes.

Methodology and Experiments

Field Measurement of Infiltration

A particular experimental site of infiltration of groundwater into sewers is in city district Prague 4 – Hrncire. The district comprises a newly constructed residential area with unfavourable hydro-pedological conditions.

The groundwater level in Prague-Hrncire is very high, often only 0.5 metres below the surface. This groundwater is the source of infiltration into the sewer system, gravitational in part, and succeeded by a vacuum sewer. A camera-assisted survey revealed infiltration of ballast water through house connections, in particular through house connections No.2 in Babockova Street (Figure 1.2), and extension No.1 behind pit No S6.

The discovery of actual points of infiltration and the plausibility of conducting a measuring of this small basin made it possible to attempt assessment of the volume of infiltrated water as bilateral ingress into horizontal drainage of the above specified extensions and compare the calculations to the measurements.

For constant flow, the infiltration per unit of length of the drainage pipe can be expressed as follows:

$$q = K \cdot S \cdot I, \tag{4}$$

where K is the factor of hydraulic flow ($m.s^{-1}$), for the given location, the value of the coefficient is approximately $K = 1.10^{-5}$ $m.s^{-1}$, S is surface area for unit pressure, corresponding to unit of length of the drainage (m^2), I is pressure gradient corresponding to constant pressure level (1,0).

The next calculation will assume that the floor of the drainage element, in our case the sewer extension, rests on an impermeable layer, that the groundwater flow q is constant at the distance of x from the house connections on both sides per length unit, and that the external source of water is in the subsoil.

The differential formula is as follows:

$$q = 2Ky\frac{dy}{dx}, \tag{5}$$

where y is water depth (m), x is length (m), $\frac{dy}{dx}$ is gradient.

Integral to the above formula is the limiting of values of x = 0 and y = depth of water in the extension (this depth is negligible). This produces the final formula for infiltration flow into the house connections:

$$Q = 2K\frac{h^2}{R}L_p, \tag{6}$$

where h is the difference between the non-lowered level of groundwater and the floor of the extension, set at 0.6 metres, R is the radius of lowering of the level due to infiltration of groundwater for the given soil set at 5 metres, L_p is the actual length of the house connections, lengths are 19.75 m and 22.40 m respectively. (Figure 1.2. – HC 1 and HC 2).

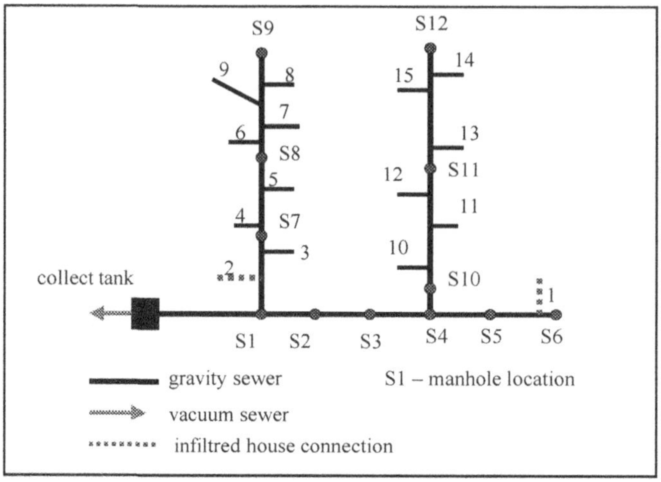

Figure 1.2 Sewer system scheme of area Prague-Hrncire.

Results and discussion

The calculation according to formula (6) yields values of infiltration $Q_1 = 1.71$ l.min^{-1} (for HC 1) and infiltration $Q_2 = 1.94$ l.min^{-1} (for HC 2). Therefore, the total infiltration from this locality is $Q_{1+2} = 3.64$ l.min^{-1}.

The value was compared to the value of the daily minimum, which can be considered as a good approximation of extension infiltration. Recorded the daily minimum flow in sewers was $Q_{Dmin} = 2.75$ l.min^{-1} (one week observation) with flow data available every ten minutes (Figure 1.3).

In fact, the difference between value $Q_{1+2} = 3.64$ l.min^{-1} obtained by method according to equation (6) and measured value $Q_{Dmin} = 2.75$ l.min^{-1} is equal to 24%. The method proposed by equations (4) ~ (6) is designed for a first approach to obtain result, principally, equation (3) is applied for a linear construction as drainage with uniform perforated pipes. Whereas, damaged house connection pipe is perforated unevenly.

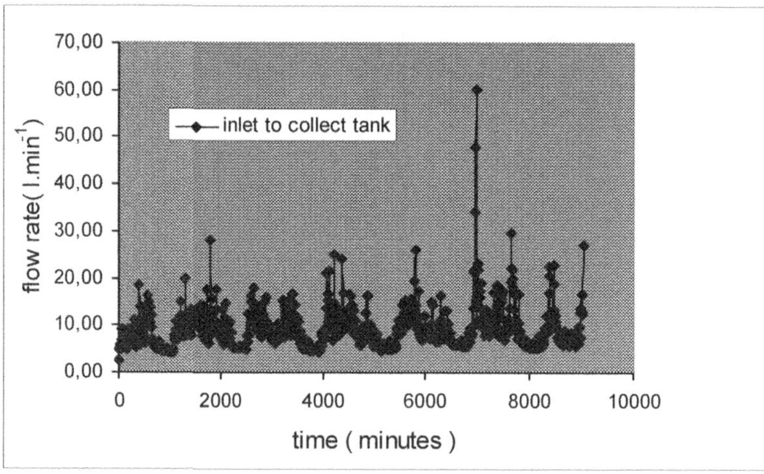

Figure 1.3 Inlet to collect tank recognized every ten minutes.

Field Measurement of Exfiltration

The most suitable location for experimental measurement of exfiltration was found in the city district Prague-Troja. Based on information from local inhabitants a particular "problem-ridden" street was selected, where house-basements and water-supply manhole frequently get flooded from house connections as well as from the main sewer. The Prague Sewer Net Administrator (PVK,a.s.) conducted a camera-aided inspection of the main sewer, including the house connections. The ascertained condition was classified as extended defects in several places, e.g. broken and loose pipe wall, gap in joint, etc.

The field measurement of exfiltration was conducted for a section of main sewer (demarcated by the inspection manhole M1 and the connecting inspection manhole M2), at a length of 46.1 metres, with ceramic pipe DN 250 mm, and with connections of two house connections DN 200 mm (Figure 1.4). The standard measurement method was applied, as in the case of water-tightness test. The experiment was commenced by blocking the lower end of the selected sewer segment with a sealing balloon at manhole M2 and filling the sewer segment with water to a specified level, followed by measuring the water level decrease at manhole M1 in time. The exfiltration rate corresponds to the lost volume of water in the given time, expressed as volume change ΔV at manhole M1 in time Δt. The resulting volume of exfiltration related to water level is indicated in (Figure 1.5). Measuring was repeated approximately after 20 minutes.

As the second part of experiment was sampling from water-supply manhole. Result of sampling analysis proved discharge of waste water from house connection to soil environment consequently to basement and manhole.

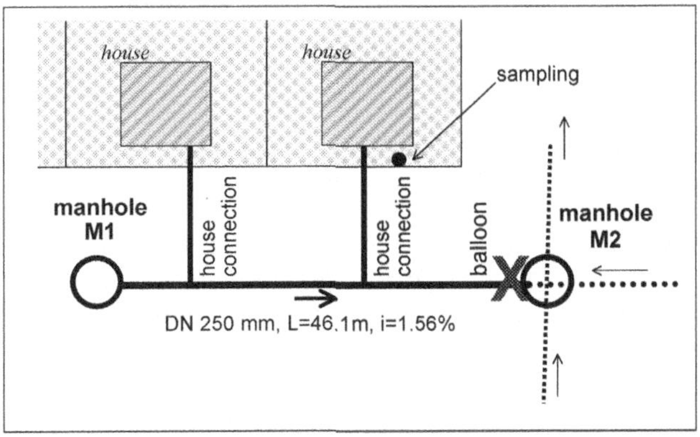

Figure 1.4 Layout of the experimental site of Prague-Troja

Results and discussion

The graph of exfiltration as shown on Figure 1.5 indicates that outflow from the piping is directly dependent on the water level in the pipe *(h)*. With decreasing water level, the exfiltration volume drops. The outflow is also directly dependent on the filled volume of the system including house connections *(V)* under various water levels. That means:

$$Exf = f(h,V), \tag{7}$$

where h is high of water level (m), V is volume of water in pipe (m^3), Exf is exfiltration amount ($m^3.s^{-1}$) or ($l.s^{-1}$).

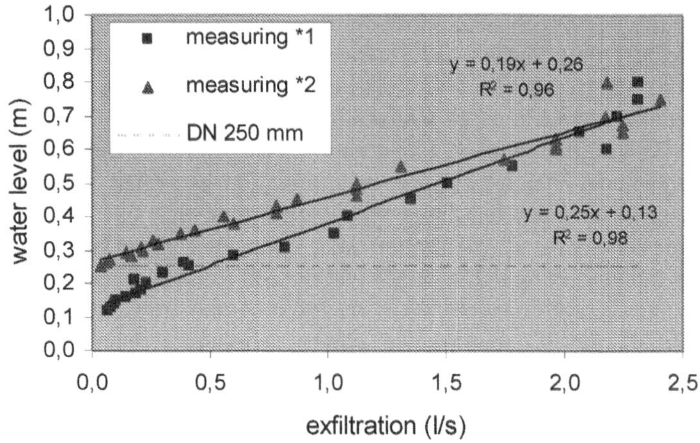

Figure 1.5 Exfiltration from sewer, observing water level in manhole M1.

In order to be able to compare measurements from various sites with different conditions (piping condition, type of backfill, etc.), the exfiltration level is expressed as a relative value, related, for example, to a length unit of piping, number of joints, or percentage of piping capacity. In the case of the experimental sewer in Prague-Troja, the following values apply: for a 75% filling of the DN 250 mm pipe the exfiltration rate is 0.23 l/s. The relative exfiltration rate related to 1 meter of piping is 0.0035 l/s, and the exfiltration rate related to the piping joint is 0.0034 l/s.

Conclusion

Infiltration assessment

The method of calculation of infiltration into the house connection is based on an analogy with infiltration into horizontal drainage, the calculation being based on the Darcy Filtration Law. In this case the result is close to values of the daily minimum flows. It must be borne in mind that application of this calculation has its limitations. In case of the calculation being applied to systemic drainage, it is assumed that the drainage piping is perforated in uniform fashion along the whole length and that the drainage system comprises parallel and equidistant drainage pipes. It must be emphasised that the points of infiltration into sewers are leaky joints and/or crevices, i.e. apertures of variable size and variable distribution along the length of the piping, while the water inflow seldom follows the Darcy Law, the sources often being scattered where the Bernoulli formula applies, as in an example of infiltration from a waterlogged cavern. Further research should concentrate on verification of potential application of the drainage theory for approximation of infiltration of groundwater into sewer systems based on knowledge of the mutual position of the sewer, the level of the groundwater, the permeability of surrounding subsoil, and the actual condition of the sewer system.

Exfiltration assessment

The performed measuring proved an undesired outflow of wastewater from the sewer (main sewer and house connections), which is damaged and in a condition that is causing ingress of water into basements and water-supply manhole.

The methodology of exfiltration measurement is based on standardised regulations applicable in the Czech Republic, regarding water tightness test of sewers. However, it has been adjusted for the conditions of damaged piping in operation. The most sensitive part of the method is in ensuring the tightness of the stopping up of the sewer with sealing balloons.

Examination of the infiltration/exfiltration issues of house connections requires the application of knowledge in the area of hydraulics, hydraulics of groundwater and transport processes in soil. It also requires execution of the taking of measurements of house connections and appropriate processing of obtained data.

Acknowledgements

This study has been carried out within the framework of the European research project APUSS (Assessing Infiltration and Exfiltration on the Performance of Urban Sewer Systems) which partners are INSA de LYON (FR), EAWAG (CH), Technical University of Dresden (DE), Faculty of Civil Engineering at University of Prague (CZ), DHI Hydroinform a.s. (CZ), Hydroprojekt a.s. (CZ), Middlesex University (UK), LNEC (PT), Emschergenossenschaft (DE) and IRSA-CNR (IT). APUSS is supported by the European Commission under the 5th Framework Programme and contributes to the implementation of the Key Action "Sustainable Management and Quality of Water" within the Energy, Environment and Sustainable Development Contract n° EVK1-CT-2000-00072.

References

WATER ENVIRONMENT FEDERATION (1999) *Control of Infiltration and Inflow in Private Building Sewer Connections.* Alexandria, VA (USA)

Cislerova, M., Vogel, T. (1998) *Transportni procesy*, CVUT Praha, (in Czech)

CSN 756101, (1994) *Stokove site a kanalizacni pripojky*, CNI Praha (in Czech)

Haloun, R. (1993) *Modelovani odtoku z intravilanu*, CVUT Praha, (in Czech)

Hlavac, V., Hanton, K., Krizek, P., Smutny, V. (2003) *Nekolik inovaci pro inspekci kanalizacniho potrubi televiznimi inspekcnimi systemy*, Sovak 6/2003, 171-173 (in Czech)

Davies, J. P., Clarke, B. A., Whiter, J. T., Cunningham, R. J. (2001) *Factors influencing the structural deterioration and collapse of rigid sewer pipes*, Urban Water 3.

Vollertsen, J., Hvitved-Jacobsen, T. (2002) *Exfiltration from gravity sewers-a pilot scale study*, Water Science and Technology, 47(4), 69-76

Ellis, J. B., Rewitt, D. M. (2002) *Sewer losses and interactions with groundwater quality*, Water Science and Technology, 43(3), 195-202

Measurement of infiltration rates in urban sewer systems: use of oxygen isotopes

J. De Bénédittis (1), J.-L. Bertrand-Krajewski (1)

(1) URGC Hydrologie Urbaine, INSA de Lyon, Bâtiment Coulomb, 34 avenue des Arts, 69621 Villeurbanne cedex, France

Abstract The paper presents the principle of a method to measure infiltration rates in sewer systems based on the use of oxygen isotopes and its application in Lyon (France). Investigations in the urban area of Lyon benefit from the isotopic differences between underground waters originating from the two rivers Rhône and Saône and from their associated alluvial aquifers. The oxygen isotopic composition of the Rhône water is roughly 3 ‰ lighter than that of the Saône river, due to the large differences in the mean altitude and topographic situation of their catchments. Large amounts of water are pumped from the Rhône aquifer for drinking water supply. As a consequence, this difference in the oxygen isotopic composition between wastewater and local groundwater can be used to evaluate infiltration in sewers in some parts of the city. The results obtained in a case study allow to evaluate the reliability and the uncertainties of this method.

Keywords Infiltration, oxygen isotopes, uncertainty, reliability, sewer systems

Introduction

Urban sewer systems constitute a very significant asset in European cities. As sewer systems are deteriorating, their functional and structural performance is impaired, leading to groundwater infiltration into sewer systems, which is particularly detrimental to the efficiency of wastewater treatment plants due to hydraulic overloading and dilution of pollutant loads. Infiltration is critical on a long-term basis for sustainable urban water management and has significant economic consequences for cities through the European Union. Infiltration measurements can also be used by sewer operators as a performance indicator relating to structural state and environmental efficiency of sewer systems. However, traditional measurement methods (see Table 1) are still subject to considerable uncertainties due to their underlying assumptions and general principles.

Two types of methods can be distinguished : i) flow rate methods (F) based on the analysis of daily hydrographs and ii) chemical methods (C) based on the analysis of the dilution of pollutants. For any method applied during dry weather periods, two principles are used either separately or jointly: (1) infiltration is calculated by substraction of a theoretic strict wastewater flow which is usually estimated from the annual drinking water consumption or by a reference value of discharge per inhabitant; (2) infiltration is supposed close to night flow. These two basic principles should be carefully discussed because i) the strict wastewater flow presents daily and seasonal variations linked to human activities, ii) the use of a reference value can induced large uncertainties and iii) the night flow is not only due to infiltration but also to all permanent contributions like groundwater pumping for

 Sewer Networks and Processes within Urban Water Systems edited by J.-L. Bertrand-Krajewski, M. Almeida, J. Matos & S. Abdul-Talib. ISBN: 1 84339 506 1.

cooling, drinking water leakages, etc. Thus, traditional methods are affected by uncertainties in both the origin of infiltrated water and in the reliability of infiltration rate estimations.

Table 1 Synthesis of traditional methods to estimate infiltration in sewers

Methods	Origin	Reference	Principle	Type
Dry weather flow	-	-	(1)	F
Dry weather flow bis	Switzerland	(Hager *et al.*, 1985)	(1)	F
Density average	Austria	(Ertl *et al.*, 2002)	(1)	F
Annen & Muller	Austria	(Annen, 1980)	(1)	F
Triangle	Germany	(Weiss *et al.*, 2002)	(1)	F
Moving minimum	Germany	(Weiss *et al.*, 2002)	(1)	F
Difference of daily flow	France	(Joannis, 1994)	(1)	F
Difference of night flow	France	(Joannis, 1994)	(2)	F
Minimum night flow	France	(Renault, 1983)	(2)	F
Corrected night flow	France	(Renault, 1983)	(2)	F
Corrected night flow bis	Switzerland	(Hager *et al.*, 1985)	(2)	F
Shape parameter	France	(Joannis, 1994)	(2)	F
IMHOFF	France	(Renault, 1983)	(1)	C
Swiss	Switzerland	(Hager *et al.*, 1985)	(1) and (2)	C
Hybrid or Horizon	France	(Horizon, 1992)	(1) and (2)	C

The study of the origin of wastewater is important because sewer operators are focused on groundwater infiltration, i.e. not only real infiltration through sewer defects, but also discharges of undeclared groundwater pumping. So it is necessary to develop a new method to estimate infiltration, which allows the identification of the origin of wastewater flow components during dry weather : drinking water and groundwater. The principle of this new method consists to use natural tracers in water, like in hydrogeology, and especially the isotopic composition of the elements in the dissolved phase or even in the water molecule itself. The use of such tracers offers two advantages: i) the identification and the quantification of wastewater flow components and ii) the determination of the uncertainty in infiltration estimations. It is assumed that each wastewater component has a specific content in tracer, which is i) significantly different from that of the other component and ii) constant whatever the conditions. It is necessary to use tracers which are conservative and not altered by phenomena like adsorption on sewer sediments, oxydo-reduction, pH, temperature variations, etc. In the specific context of sewer systems, the analysis of oxygen isotopic composition, usually expressed as $\delta^{18}O$, seems to be an interesting approach. This idea to use water isotopes to measure infiltration in sewer systems has been first proposed by W. Gujer (EAWAG, Switzerland) as one of the key ideas on which the European research project APUSS (Assessing Infiltration and Exfiltration on the Performance of Urban Sewer Systems) is based. Some brief information about the method and its application has already been given in Kracht *et al.* (2003). The main objective of this paper is to present i) the measurement method by use of oxygen isotopes, i.e. the $\delta^{18}O$ method, ii) the investigations carried out in Lyon, France in order to check its applicability and iii) the results of a case study in order to check its reliability.

Principle of the $\delta^{18}O$ method

Definition of $\delta^{18}O$

The water molecule is constituted by two elements, O and H, each one possessing three stable or radioactive isotopes. The relative abundance of an isotope in natural water depends on several factors : the location of rain events (altitude, latitude, distance from the ocean), the cycle of evaporation/condensation, exchanges with minerals, etc. The relative abundance of the stable isotope ^{18}O is expressed according to its abundance in the standard mean oceanic water (SMOW). The isotopic ratio $^{18}O/^{16}O$ in water samples, also named $\delta^{18}O$, is described by its relative variations compared to SMOW, expressed in ‰ :

$$(\delta^{18}O)_{sample} = \frac{(^{18}O/^{16}O)_{sample} - (^{18}O/^{16}O)_{SMOW}}{(^{18}O/^{16}O)_{SMOW}} * 1000 \qquad \text{Eq. 1}$$

Preparation of water samples

The measurement of $\delta^{18}O$ requires only 2 or 3 mL of water : so it is not necessary to sample large volumes of water. Raw samples are taken in 500 mL or 1 L PVC bottles. They are then filtered in conical filters and sent to the laboratory in 60 mL glass flasks which are completely filled with water to avoid any contact with atmosphere and any evaporation

Application of $\delta^{18}O$ measurements to estimate infiltration

The estimation of infiltration by the use of oxygen isotopic composition is analogous to the method of decomposition of flood hydrographs developed in hydrogeology (Blavoux, 1978). In dry weather, the mixing of strict wastewater flow Q_{WW} which origin is drinking water with infiltration flow Q_{INF} which origin is groundwater constitutes the total wastewater flow Q_T :

$$Q_T = Q_{WW} + Q_{INF} \qquad \text{Eq. 2}$$

At catchment scale, infiltration is assumed to result only from groundwater infiltration and strict wastewater is assumed to result only from drinking water consumption. The $\delta^{18}O$ of groundwater samples (or other possible sources of infiltration as rivers, creeks, etc) constitutes the reference value δ_{INF} for infiltration components. The $\delta^{18}O$ of drinking water samples constitutes the reference value δ_{WW} for strict wastewater components. The $\delta^{18}O$ method can only be applied if δ_{WW} is significantly different from δ_{INF}, which is problematic if drinking water and potentially infiltrated groundwater originate from the same aquifer. The applicability of the $\delta^{18}O$ method depends on the hydrogeological context and on the location of drinking water pumping stations. Moreover, reference values must be constant at the space and time scales of the study. If these conditions are satisfied, the measurement of $\delta^{18}O$ in total wastewater δ_T, in drinking water δ_{WW} and in potential infiltration water δ_{INF} sampled simultaneously in the catchment allows to evaluate the respective proportions of these two components in the total wastewater flow, according to a set of mixing equations :

$$Q_T\,\delta_T = Q_{WW}\,\delta_{WW} + Q_{INF}\,\delta_{INF} \qquad \text{Eq. 3} \qquad\qquad \delta_T = a\delta_{WW} + b\delta_{INF} \qquad \text{Eq. 4}$$

$$a = 100\frac{Q_{WW}}{Q_T} \quad \text{Eq. 5} \qquad b = 100\frac{Q_{INF}}{Q_T} \quad \text{Eq. 6} \qquad a + b = 100 \quad \text{Eq. 7}$$

The variables a and b expressed in % are respectively defined as the fraction of drinking water (or strict wastewater) and the fraction of groundwater (or infiltration fraction) in the total wastewater flow Q_T. Instantaneous, mean daily or night and diurnal mean samples can be used, depending on the objectives of the study. The two most interesting variables are the infiltration fraction b and the infiltration flow rate Q_{INF}, which are defined by :

$$b = 100\frac{\delta_T - \delta_{WW}}{\delta_{INF} - \delta_{WW}} \quad \text{Eq. 8} \qquad\qquad Q_{INF} = bQ_T \quad \text{Eq. 9}$$

Uncertainty in infiltration estimation

The application of the law of propagation of uncertainties (NF ENV 13005, 1999) to Eq. 8 allows to calculate the uncertainty Δb (95 % confidence interval) in the fraction b :

$$\Delta b = \frac{100\Delta\delta\sqrt{2}}{\delta_{INF} - \delta_{WW}}\sqrt{b^2 - b + 1} \qquad \text{Eq. 10}$$

The uncertainty in b depends on the uncertainty in the $\delta^{18}O$ analyses, on the difference of $\delta^{18}O$ between drinking water and infiltration water and on the value of b itself. Considering that the mean value of the uncertainty in $\delta^{18}O$ analyses is 0,1 ‰, one can draw various curves of uncertainty in b according to the theoretic difference observed between the reference values δ_{INF} and δ_{WW}. The uncertainty in b is lower when b is closed to 0,5 and when the difference between δ_{INF} and δ_{WW} increases. This last factor is the most influencing one. A proper estimation of the uncertainty is important because it is necessary to draw valuable conclusions regarding infiltration. An estimation of the infiltration is considered as pertinent only if the relative uncertainty $\Delta b/b$ is lower than 1. If $\Delta b/b$ is higher than 1 (grey area in Figure 1), the $\delta^{18}O$ method is considered as not adequate. The intersections between the dotted curve and the uncertainty curves are used to define the threshold corresponding to measurable infiltration fractions. The application of the law of propagation of uncertainties to Eq. 9 allows to calculate the uncertainty ΔQ_{INF} in the calculated value of Q_{INF} (sampling uncertainties are neglected in this calculation) :

$$\Delta Q_{INF} = 0.01\sqrt{\Delta b^2 Q_T^{\;2} + \Delta Q_T^{\;2} b^2} \qquad \text{Eq. 11}$$

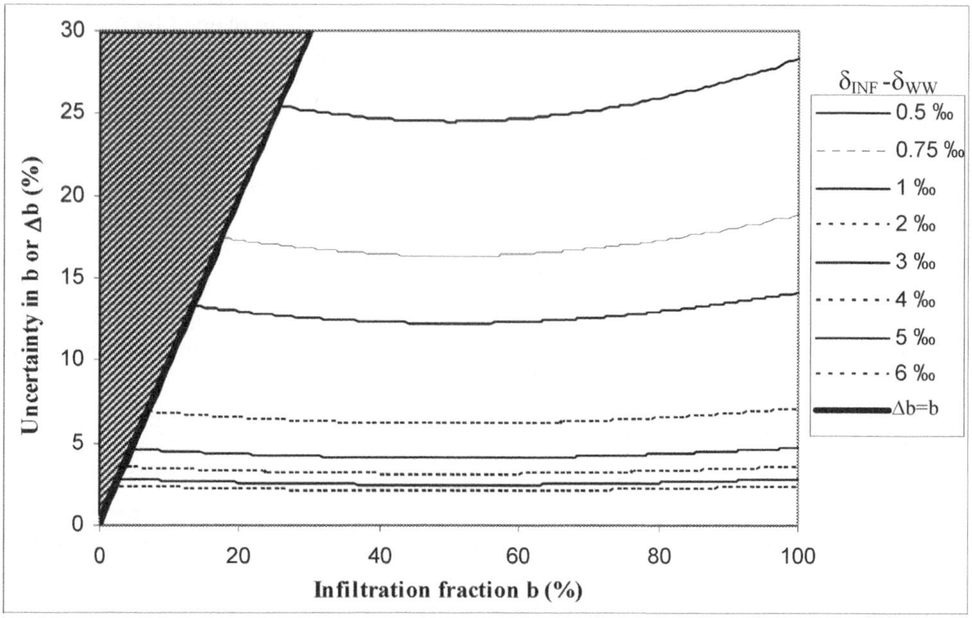

Figure 1 Curves of uncertainty in b.

Applicability of the $\delta^{18}O$ method in Lyon

Preliminary measurement campaigns are necessary to test the applicability of the $\delta^{18}O$ method. The objective is to highlight significant $\delta^{18}O$ differences between drinking water and groundwater during dry weather periods. Drinking water in Lyon is pumped from the aquifer in the Rhône modern alluvia. A sample taken at the production plant which supplies all the city of Lyon was analysed. Its $\delta^{18}O$ constitutes the reference value δ_{WW}. Infiltration water in the sewer system of Lyon during dry weather periods can have two major origins : the Rhône river, the Saône river and their respective alluvial aquifers. Samples were taken in both rivers. Their $\delta^{18}O$ constitute the reference values δ_{INF}. Two sampling campaigns have been carried out in March and September 2002 in order to evaluate seasonal effects on the reference values. The results are given in Table 2.

Table 2 $\delta^{18}O$ of the main components of total wastewater flow in Lyon

Sample description	March 2002		September 2002	
	$\delta^{18}O$ (‰)	$\|\delta_{INF} - \delta_{WW}\|$	$\delta^{18}O$ (‰)	$\|\delta_{INF} - \delta_{WW}\|$
Drinking water	-9,44	-	-10,75	-
Rhone river	-11,05	1,61	-10,53	0,22
Alluvial aquifer of Rhone	-10,81	1,37	-10,47	0,28
Saône river	-8,11	1,33	-7,45	3,3
Alluvial aquifer of Saône	-8,03	1,41	-7,32	3,43

In March 2002, an average variation of 1,5 ‰ between drinking water and all possible sources of infiltration is observed. The $\delta^{18}O$ method can be applied and, according to Fig. 1, the infiltration fraction is meaningful for *b* values higher than 9 %. In September 2002, there is no clear distinction between drinking water and parasitic water which origin is the Rhône river or the groundwater from its alluvial aquifer. This fact can be explained by the seasonal variations of the Rhône river and of groundwater levels. The drinking water is pumped from the Rhône alluvial aquifer after artificial re-injection. As the level of the Rhône river is higher during the summer because of snow melting in the Alps where the Rhône has its source, the groundwater drains the Rhône and δ_{WW} is close to the $\delta^{18}O$ of the Rhône river. In winter and spring, the Rhône level is lower, it drains the groundwater and δ_{WW} is more influenced by local groundwater than by the Rhône. As a consequence, the $\delta^{18}O$ method can not be applied during all the year to sewers located close to the Rhône river and to its alluvial aquifer. On the contrary, an average variation of 3 ‰ between δ_{WW} and infiltration from the Saône river or its alluvial aquifer is observed. The $\delta^{18}O$ method can be applied through the year. According to Fig. 1, the infiltration fraction is meaningful for values higher than 4,5 %.

Reliability of the $\delta^{18}O$ method : a case study in Lyon

The catchment of Ecully in Lyon has an area of 245 ha and a residential land use. Continuous measurements of wastewater flow rate and pollutants concentrations are available since April 2001. The application of the traditional methods listed in Table 1 reveals that infiltration is significant. The $\delta^{18}O$ method has been applied during one day, with 24 mean hourly samples of wastewater taken between 12/03/2003 at 10:00 and 13/03/2003 at 10:00. The reference values δ_{INF} and δ_{WW} are provided by two instantaneous samples taken respectively in the drinking water network and in a creek located closed to the sewer system. All measurement results are given in Figure 2. A variation of 1,4 ‰ is observed between the reference values for strict wastewater and groundwater : the $\delta^{18}O$ method is applicable. As all values of δ_T are between the reference values δ_{INF} and δ_{WW}, one can conclude that the total wastewater is a mixture of groundwater and drinking water (if no other sources are considered). This leads to positive infiltration rates, which confirms the reliability of the method. Moreover, during the night period, the values of δ_T tend to δ_{INF} which corresponds to a logical reduction of domestic discharges. The proportion of the two components can be calculated for every mean hourly sample of wastewater in order to obtain the composition of the total daily hydrograph (Figure 3).

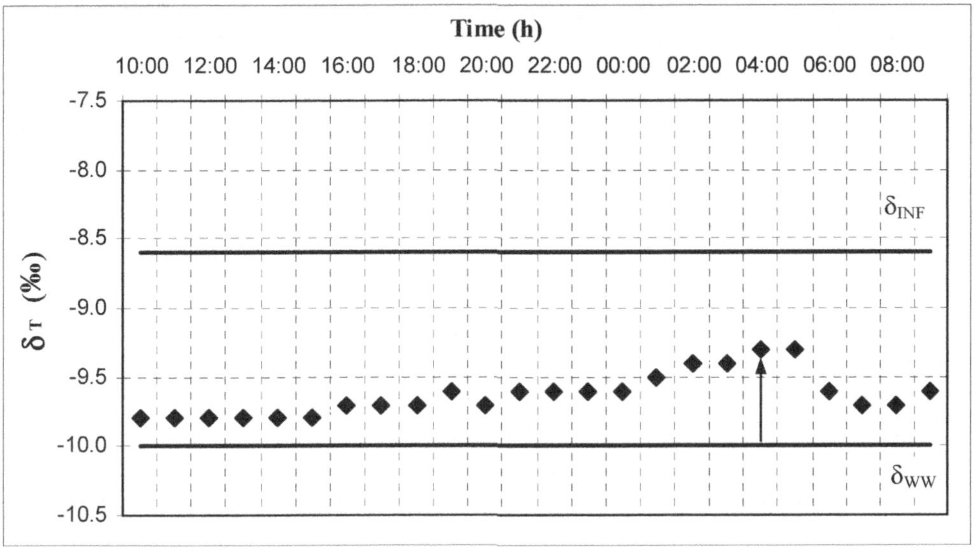

Figure 2 Values of $\delta^{18}O$ measured in the Ecully catchment

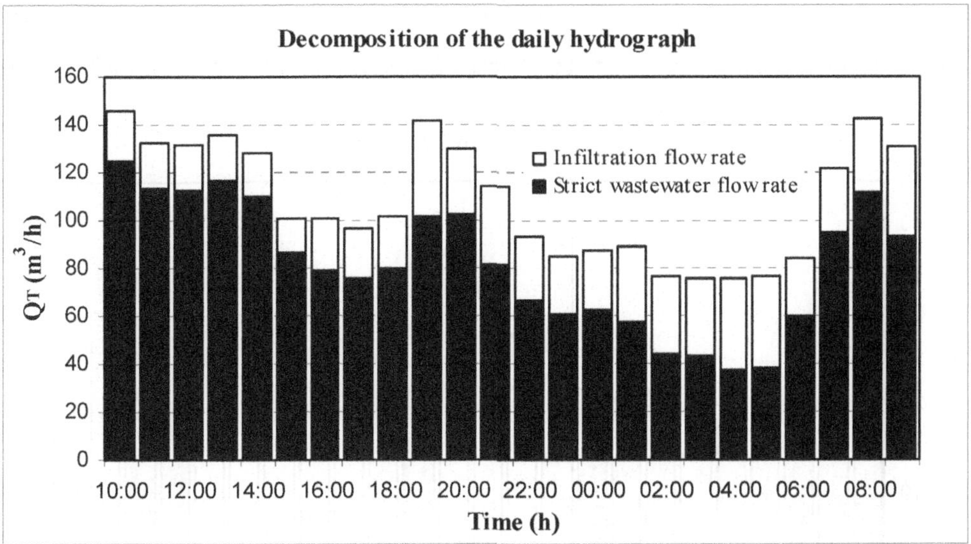

Figure 3 Composition of a daily total hydrograph using the $\delta^{18}O$ method

The reliability of the method means its capacity to reproduce the characteristics of a daily hydrograph with the presence of infiltration. Usually, infiltration flow is assumed to be constant at daily scale while strict wastewater flow is characterised by strong variations due to human activities with two peak flows in the morning and in the evening. These properties can be partly observed in Figure 3, and appear more clearly with the hourly variations of the infiltration fraction b and of the total wastewater flow Q_T as shown in Figure 4. The hourly variations of the infiltration fraction *b* are inversely proportional to the hourly variations of the total wastewater flow Q_T. The $\delta^{18}O$ method reproduces the daily cycle of strict wastewater with a strong decrease of the contributions during the night period represented by a strong increase of the infiltration fraction *b*. The flow peaks of the strict wastewater contribution are also characterized by a strong decrease of the infiltration fraction *b* during the evening (20:00) and the morning (07:00-09:00). From the total hydrograph given in Figure 3, one can easily derive the hydrograph of infiltration given in Figure 5.

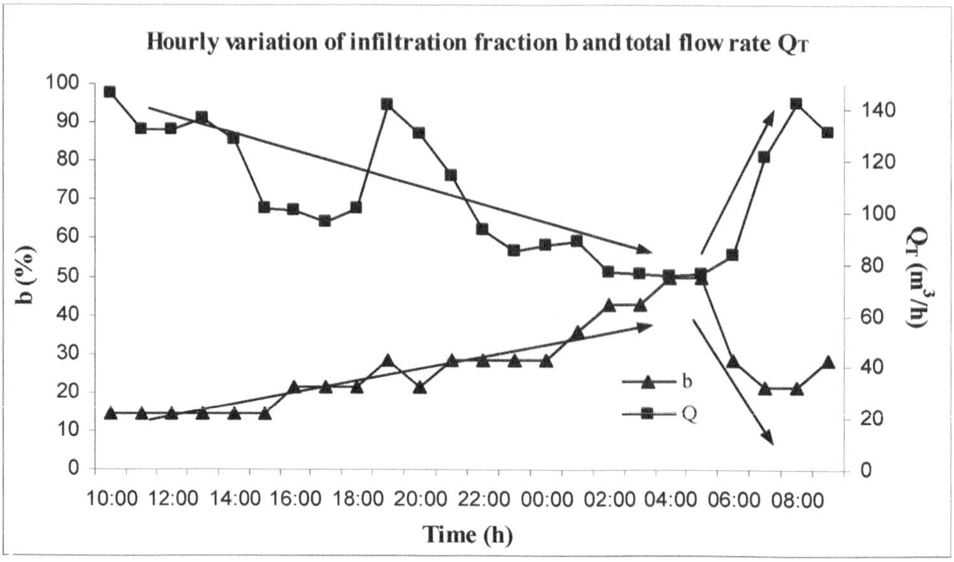

Figure 4 Hourly variations of the infiltration fraction *b* correlated with total wastewater flow Q_T

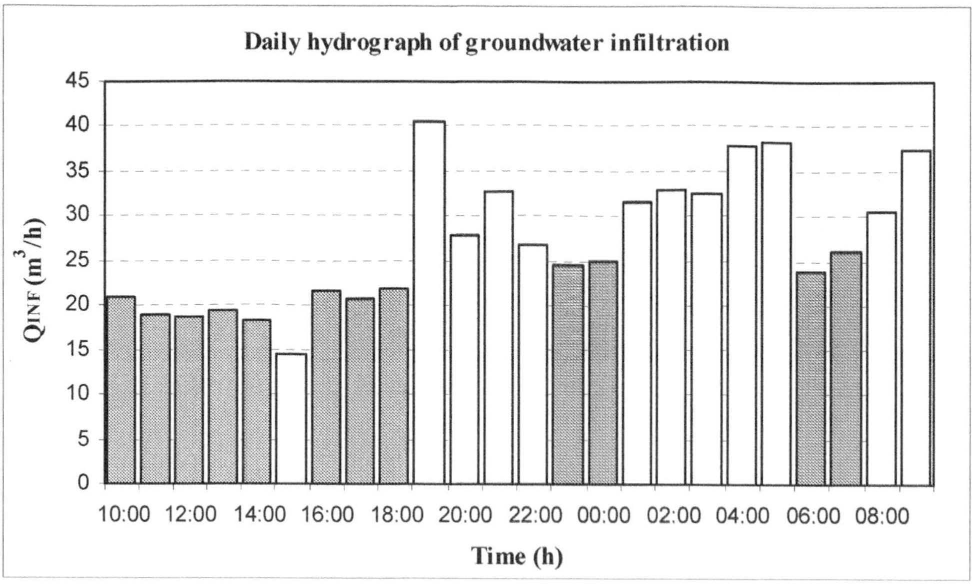

Figure 5 Daily hydrograph of groundwater infiltration

The hydrograph of infiltration shows strong variations during the day : infiltration can vary with a factor 1 to 2 during three peak periods marked with circles in Figure 5. A lower value is observed from 15:00 till 16:00. Without these peaks, the infiltration rate would be relatively constant (grey blocks). The assumption of a permanent and constant infiltration rate during the day can be questioned. The observed variations could be explained by two phenomena. The first phenomenon relates to the $\delta^{18}O$ method which provides a value of groundwater infiltration rate without knowing its way to enter the sewer system : either true infiltration trough tightness defects or discharges of groundwater pumped for various purposes (cooling, etc.). Infiltration variations could be explained by the variability of groundwater pumping contributions. Indeed, two peaks (1 and 3) are observed during the period of strict wastewater peaks. These two peaks could be explained by the contribution of groundwater pumping used specifically for domestic activities. The word specifically is used because some groundwater pumping (cooling) constitute permanent contributions. The breakdown of such permanent groundwater pumping could explain the infiltration decrease observed from 15:00 till 16:00.

The second phenomenon relates to the mechanisms regarding true infiltration. Indeed, true infiltration occurs mainly around the tightness defects which are located along the pipe wall between the wastewater surface level and the groundwater level around the sewer. When the wet perimeter in the pipe decreases during the night period (i.e. decreasing water level in the pipe), two mechanisms may simultaneously occur: i) more defects can potentially contribute to infiltration, and ii) the infiltration flow through defects which also contribute during the day period can increase. These mechanisms could explain the peak nr 2 observed from 01:00 till 6:00, mainly from 04:00 till 06H00, period during which the total wastewater flow is minimum. However, before confirming such assumptions, it is necessary to take into account the uncertainty in hourly flow rates (Figure 6).

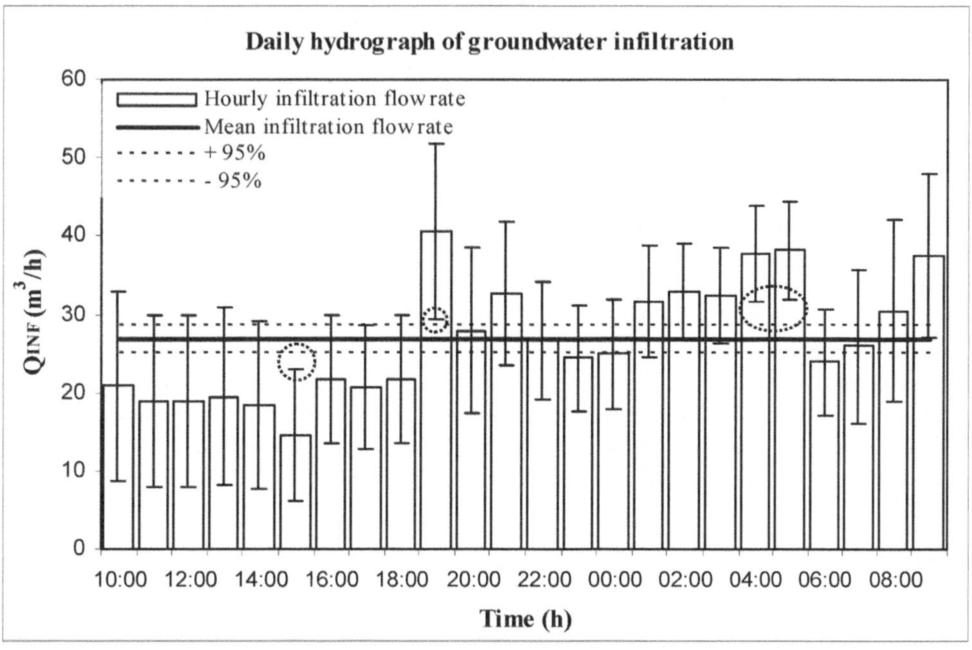

Figure 6 Uncertainties in the infiltration hydrograph

Accounting for uncertainties, all estimated values of the hourly infiltration rate are considered as valid. The low value from 15:00 to 16:00 is significantly lower than the mean value, while the peak values between 18:00 and 19:00 and between 04:00 and 06:00 are significantly higher than the mean value: the limits of the 95 % confidence intervals of the hourly infiltration rates do not cross the limits of the 95 % confidence interval of the mean daily infiltration rate. This observation may confirm the previous assumptions and also shows that the $\delta^{18}O$ method can detect transitory groundwater contributions.

Conclusions and perspectives

This study shows that the use of oxygen isotopes for the measurement of groundwater infiltration is possible if the conditions for its applicability are satisfied. The results obtained in the case study of Ecully show that the groundwater infiltration rate is variable during the day and that the $\delta^{18}O$ method can be used to study transitory contributions. A further step in the use of the $\delta^{18}O$ method will be the decrease of uncertainty in infiltration rate estimation and the analyse of the spatial variability of δ_{INF} and δ_{WW} at catchment scale and its impact on the reliability of infiltration estimations. As uncertainty depends mainly on the difference between δ_{INF} and δ_{WW}, one solution could be the use of hydrogen isotopes which are more subject to fractionation processes than oxygen isotopes because of their lighter mass. Moreover, the simultaneous use of H and O can allow to detect and to quantify three potential sources of water in the sewer system. Another way to reduce the uncertainty would consist to develop an improved sampling strategy by accounting the fact that uncertainty decreases during night periods because i) infiltration fraction is higher and ii) if several samples are taken, the uncertainty in the mean infiltration rate is lower, as shown in Figure 6. Another perspective of application would be the analysis of the location of infiltration in diagnostic studies of sewer systems. Compared to the traditional methods, the $\delta^{18}O$ method can provides information on the infiltration fraction without flow or pollutants measurements, only with instantaneous samples of wastewater, drinking water and groundwater. This method is less expensive and provides results faster.

Acknowledgements

This study has been carried out within the framework of the European research project APUSS (Assessing Infiltration and Exfiltration on the Performance of Urban Sewer Systems) which partners are INSA de LYON (FR), EAWAG (CH), Technical University of Dresden (DE), Faculty of Civil Engineering at University of Prague (CZ), DHI Hydroinform a.s. (CZ), Hydroprojekt a.s. (CZ), Middlesex University (UK), LNEC (PT), Emschergenossenschaft (DE) and IRSA-CNR (IT). APUSS is supported by the European Commission under the 5th Framework Programme and contributes to the implementation of the Key Action "Sustainable Management and Quality of Water" within the Energy, Environment and Sustainable Development Contract n° EVK1-CT-2000- 00072.

References

Annen G.W. (1980). Trockenwetterabfluss und Jahresschmutzwassermenge. Korrespondenz Abwasser, 27(6), 411-413.

Blavoux B. (1978). Etude du cycle de l'eau au moyen de l'oxygène 18 et du tritium : possibilités et limites de la méthodes des isotopes du milieu en hydrologie de la zone tempérée. Thèse de Doctorat d'Etat ès-Sciences Naturelles : Université Pierre et Marie Curie, Paris 6. 333 p.

Ertl T.W., Dlauhy F., Haberl L. (2002). Investigations of the amount of infiltration inflow in to a sewage system. Proceedings of the 3rd "Sewer Processes and Networks" International Conference, Paris, France, 15-17 April 2002, 8 p.

Hager W.H., Bretscher U., Raymann B. (1985). Methoden zur indirekten Fremdwasserermittlung in Abwassersystemen. Gas-Wasser-Abwasser, (64)7, 450-461.

Horizon (1992). Rapport d'étude diagnostic du bassin versant de la station d'épuration de Pierre Bénite. Direction de l'eau du Grand Lyon.

Joannis C. (1994). Précision des estimations de débits d'eaux parasites d'infiltration dans les réseaux de collecte des eaux usées. Influence des erreurs de mesure et de la méthode de calcul. Bulletin de Liaison des Laboratoires des Ponts et Chaussées, Janvier-Février 1994, 71-87.

Kracht O., Gresch M., de Bénédittis J., Prigiobbe V., Gujer W. (2003). Stable isotopes of water as a natural tracer for infiltration into urban sewer systems. Geophysical Research Abstracts, 5, 2 p.

Renault D. (1983). Les eaux claires parasites dans les réseaux d'assainissement (réseaux unitaires). Diagnostic et réhabilitation des réseaux. Impact et économie dans le traitement. TSM, 547-555 et 577-582.

Weiss G., Brombach H., Haller B. (2002). Infiltration and inflow in combined sewer systems: long term analysis. Water Science and Technology, 45(7), 11-19.

NF ENV 13005 (1999). Guide pour l'expression de l'incertitude de mesure. Paris (France) : AFNOR, août 1999, 113 p.

Exfiltration from Sewers Pipes – Effects of Flowrates, Different Leakage Area and Soil Beddings

M.A. Ab-Wahab*, S. Abdul-Talib*, B. Baharom* and M.S. Marwi*

*Faculty of Civil Engineering, Universiti Teknologi MARA, 40450 Shah Alam, Malaysia

Abstract This research was conducted on exfiltration of wastewater from gravity sewers. Exfiltration through different types of sewer leaks and soil beddings were studied. With the different size of leakage areas and a constant thickness of soil bedding, it was found that the exfiltration rate reduces and became constant after several days. This is also true for exfiltration from pipes with the same leakage area but on soil bedding of different thickness. By knowing the exfiltration rate, the time taken for the polluted water exfiltrated from sewer pipes to reach the groundwater can be determined so that measures can be taken to prevent it. From the experiment, it can be shown that the increased in the clogging zone resulted in the decreased of exfiltration rate due to the accumulation of organic matters at the edge of the leakage area and trapped in the pores of the bedding soil.

Keywords Exfiltration, sewerage systems, sewer leakages, wastewater, gravity sewers, clogging zone

Introduction

Sewerage system plays an important role in ensuring public health, environmental protection and enhancing the standard of living of the general population. Vollertsen and Hvitved-Jacobsen (2002) conducted pilot scale experiments on exfiltration of wastewater from gravity sewers. The effect of storm events, flushing of pipes and alternating infiltration or exfiltration were simulated in the experiments. Exfiltration through different types of sewer leaks and into different soils and grading under various flows condition were also studied.

The structural integrity of many sewer systems within urban areas has been deteriorating over many decades of operation as a result of inadequate maintenance (Davies *et al.*, 2001; Ellis, 2001; IWK, 1998). Sewer systems contain leaks, which are manifested by infiltration. Infiltration has been studied extensively and the negative effects on the operation of the urban wastewater system are well documented (Ellis, 2001).

It is reasonable to assume that when groundwater can infiltrate into sewer systems, wastewater may also exfiltrate from sewer systems under conditions where groundwater table is low. Very little is known about exfiltration mechanisms and the quantification of exfiltration wastewater and pollutants (Vollertsen and Hvitved-Jacobsen, 2002). A rather limited number of studies are available on exfiltration of wastewater from sewers and the consequences on groundwater quality. A few of these studies attempt to quantify the exfiltration based on the urban groundwater pollution (Vollertsen and Hvitved-Jacobsen, 2002). However, difficulties arise in distinguishing pollution from sewers and from other

 Sewer Networks and Processes within Urban Water Systems edited by J.-L. Bertrand-Krajewski, M. Almeida, J. Matos & S. Abdul-Talib. ISBN: 1 84339 506 1.

sources. These studies were based either on indirect determination or tracer based in-sewer flow measurements. Nevertheless, a common observation was reported, *i.e.*, exfiltration from sewers is significant. Exfiltration from sewer pipes due to leakages, though not well quantified, have been estimated to be in the range of 1 to 10 m^3/km/yr (Ellis and Revitt, 2002).

Studies on sewers in Germany have estimated the net exfiltration from sewers at some 34 m^3/km/yr (Ellis *et al.*, 2002). In a separate study by Decker and Risse (1993), average exfiltration for German sewers were reported to be much lower, *i.e.*, 1-2 m^3/km/yr. Jansen and Madsen (1996) (in Vollertsen and Hvitved-Jacobsen, 2002) reported that for a small urban catchment, exfiltration caused a water loss of up to 25%. In intercepting sewers, Knudsen *et al.* (1996) (in Vollertsen and Hvitved-Jacobsen, 2002) determined mean exfiltration rates of 3 l/km/s.

Indirect determination of exfiltration has been based mainly on the water balance for a catchment. This method is believed to yield rather rough estimates because the determination of individual contributions to the water balance involved significant uncertainties. Direct measurement of exfiltration using tracers for flow measurement in principle is a better approach (Knudsen *et al.*, 1996). However, measurement of flow in real sewers is a difficult task, especially when addressing small differences in flow. Consequently, reasonable accuracy is difficult to achieve and the results of these studies must be interpreted with care (Vollertsen and Hvitved-Jacobsen, 2002).

On the use of TV-inspection for evaluating pollution due to leakages from sewers, available literature agree that the method is inconclusive when identifying if –and to what degree – wastewater leaking is from sewers (Eiswirth and Hotzl, 1994; Knudsen *et al.*, 1996). In other words, not all damages of sewers are also leaks from exfiltration. The goodness of this postulate is difficult to assess as a missing correlation between the damages found by TV inspection and exfiltration could also be due to difficulties in determination of exfiltration.

The main objective of this study is to understand the exfiltration mechanisms from sewers, so that preventive and corrective measures can be introduced, thus reducing the risk of land and groundwater from being continuously polluted. The study includes the establishment of an experimental procedures for determining exfiltration from sewers and to determine the rate and amount of exfiltration from sewers due to different causes of leakage under various flow and soil bedding conditions.

Experimental set-up

Figure 1 shows the layout plan for the equipment used for the study of exfiltration from various pipes. The equipment was set-up near an operational aeration settling tank at Univesiti Teknologi MARA main campus in Shah Alam. This circular aeration tank has a capacity to process wastewater produced by approximately 3000 residence and non-residence populations. A supply of fresh wastewater from the aeration tank was pumped into a 150-gallon tank as shown in figure 1. The wastewater was allowed to flow into a 100-gallon tank before it flows under gravity, through four sets of 110mm PVC pipes placed at a gradient of between 1.0 to 1.5% . The flow rate of 0.3 m/s were maintained by controlling the inlet valves.

Four 300mm diameter by 750 mm height column reactors (marked A, B, C and D in figure 1) were placed under the pipes. Figure 2 and Figure 3 show the cross-sections of one of the column reactor. A stainless steel net with 1mm openings was placed at the bottom of the column and two layers of soil were laid on top of it consisting of a layer of gravel underneath a layer of sand. The gravel was used to support and control the thickness of the sand bedding and performed as an underdrain. Different height of sand bedding can be simulated by varying the thickness of the gravel and sand layers.

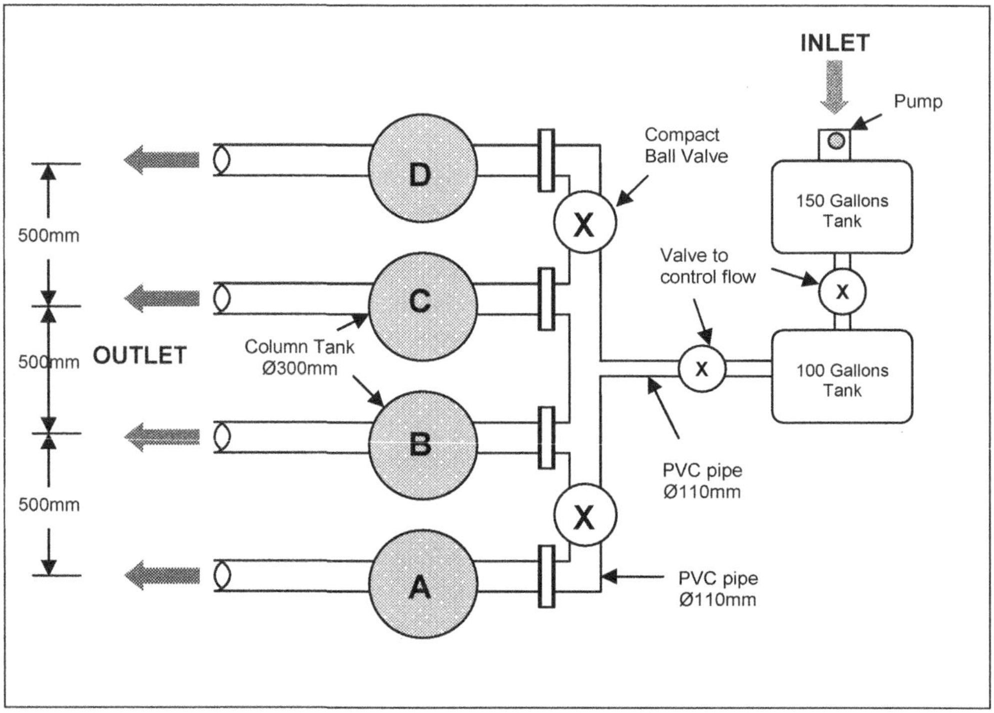

Figure 1 Experimental Set-up for Exfiltration Experiment

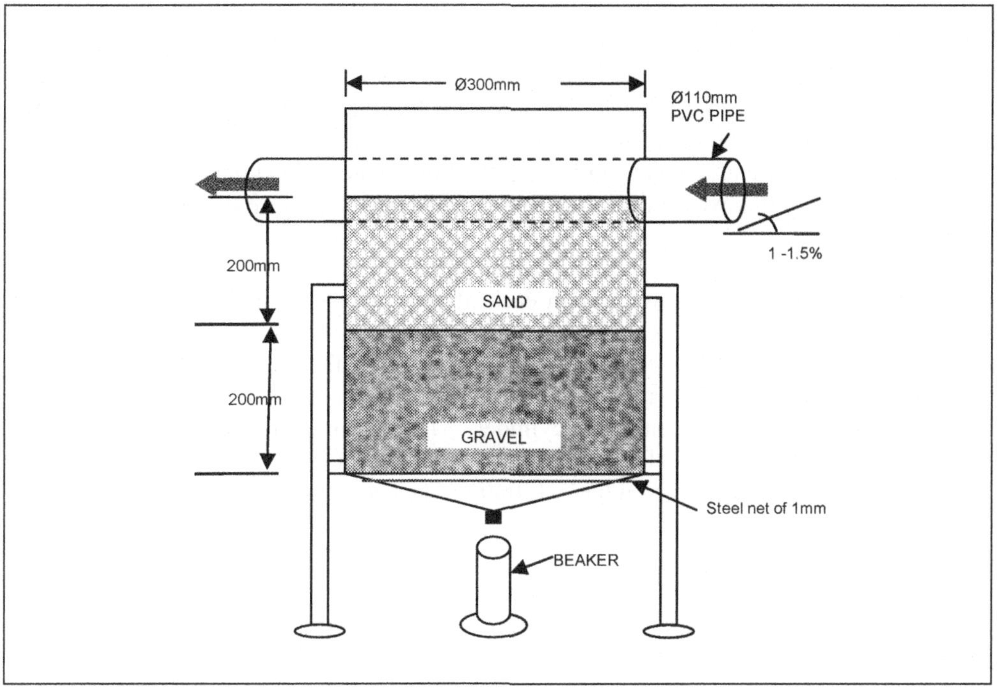

Figure 2 Cross-section of the column tank

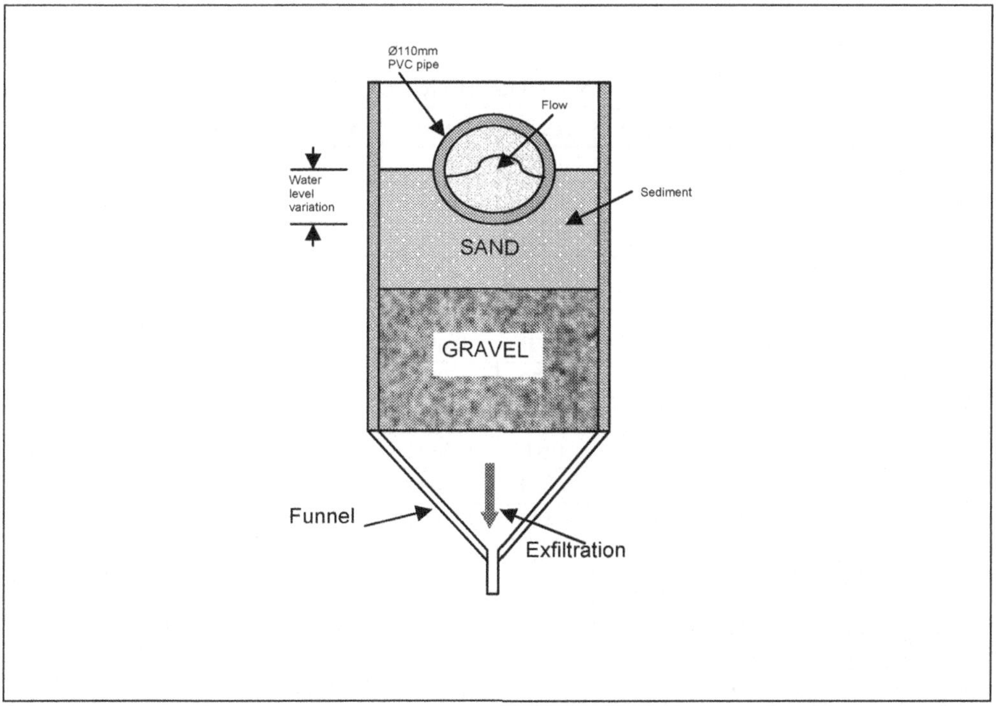

Figure 3 Wastewater flow variations

Experimental works

During the experiments, all the PVC pipes and the column reactors described above were used. A hole is made underneath each pipe above the reactor. Different size of hole is made in each pipe to simulate the different sizes of leakage. In this study, three different sizes of leakage were used; 10 cm^2, 20 cm^2 and 30 cm^2. Sand bedding thicknesses of 200mm and 300mm were used in the experiments and five (5) different tests were made per each of the set-ups.

Wasterwater was allowed to flow through each of the PVC pipes at a constant rate of 0.3 m/s. The excess wastewater was collected and discharged back into the aeration tank at the outlet. Exfiltration from the leak in the pipe was allowed to flow through the sand bedding an into the free-draining gravel layer. A funnel underneath a column reactor then channeled the exfiltrated water into a measuring cylinder. The volume of water collected in the measuring cylinder was recorded with time. The exfiltration rate through the soil bedding were measured for several days until the exfiltration rate became constant.

Results and discussions

Table 1 and Table 2 and Figure 4 below show the results of the exfiltration rate through a leak of a constant size (20 cm^2) but in soil bedding of different thickness (200 mm and 300 mm).

Table 1 Exfiltration Rate Under leak of 20 cm^2 and Soil Bedding Thickness of 200 mm at column reactor A

Time (min)	Exfiltration rate (Litre/day) x 10^{-4}							
	Test 1	Test 2	Test 3	Test 4	Test 5	**Average**	min	max
0	29.31	47.2	69.4	1.74	3.47	**30.22**	1.74	69.4
5	18.13	1.6	9.03	2.08	7.64	**7.70**	1.6	18.13
10	6.94	14.38	13.89	1.67	9.72	**9.32**	1.67	14.38
30	4.9	16.67	1.04	0.694	3.75	**5.41**	0.694	16.67
60	2.85	14.03	1.53	1.11	0.694	**4.04**	1.11	14.04
90	2.05	13.54	1.18	0.972	0.3472	**3.62**	0.3472	13.54
120	1.25	13.89	1.04	1.25	0.0694	**3.50**	0.0694	13.89
180	1.04	11.81	0.694	1.04	0.0694	**2.93**	0.0694	11.81
420	0.764	9.38	0.694	0.694	0.2083	**2.35**	0.2083	9.38
660	0.521	2.71	0.486	0.486	0.0694	**0.85**	0.0694	2.71
900	0.347	1.32	0.417	0.347	0.0347	**0.49**	0.0347	1.32
1140	0.278	1.11	0.347	0.139	0	**0.37**	0	1.11
1440	0.278	0.903	0.139	0.347	0.0139	**0.34**	0.0139	0.903
2880	0.208	0.833	0.139	0	0	**0.24**	0	0.833

Table 2 Exfiltration Rate Under leak of 20 cm^2 and Soil Bedding Thickness of 300 mm at column reactor B

Time (min)	Exfiltration rate (Litre/day) x 10^{-4}							
	Test 1	Test 2	Test 3	Test 4	Test 5	**Average**	min	max
0	2.64	6.25	10.42	0.556	1.04	**4.18**	0.556	10.42
5	1.6	2.08	15.97	0.694	2.01	**4.47**	0.694	15.97
10	0.903	16.32	1.74	1.46	0.417	**4.17**	0.417	16.32
30	0.542	17.36	1.81	0.729	0.0694	**4.10**	0.0694	17.36
60	0.139	16.67	1.39	0.139	0.0694	**3.68**	0.0694	16.67
90	0.139	8.33	1.18	0.139	0.347	**2.03**	0.139	8.33
120	0.139	15.97	0.903	0.139	0.63	**3.56**	0.139	15.97
180	0.139	13.19	0.833	0.139	0.486	**2.96**	0.139	13.19
420	0.0694	9.72	0.486	0.278	0.0139	**2.11**	0.0139	9.72
660	0.0694	2.36	0.451	0.278	0.0139	**0.63**	0.0139	2.36
900	0.0347	1.32	0.347	0.0694	0	**0.35**	0	1.32
1140	0.0347	1.11	0.139	0.104	0.0139	**0.28**	0.0139	1.11
1440	0.0208	0.833	0.104	0.0694	0	**0.21**	0	0.833
2880	0	0.625	0.104	0.0347	0	**0.15**	0	0.625

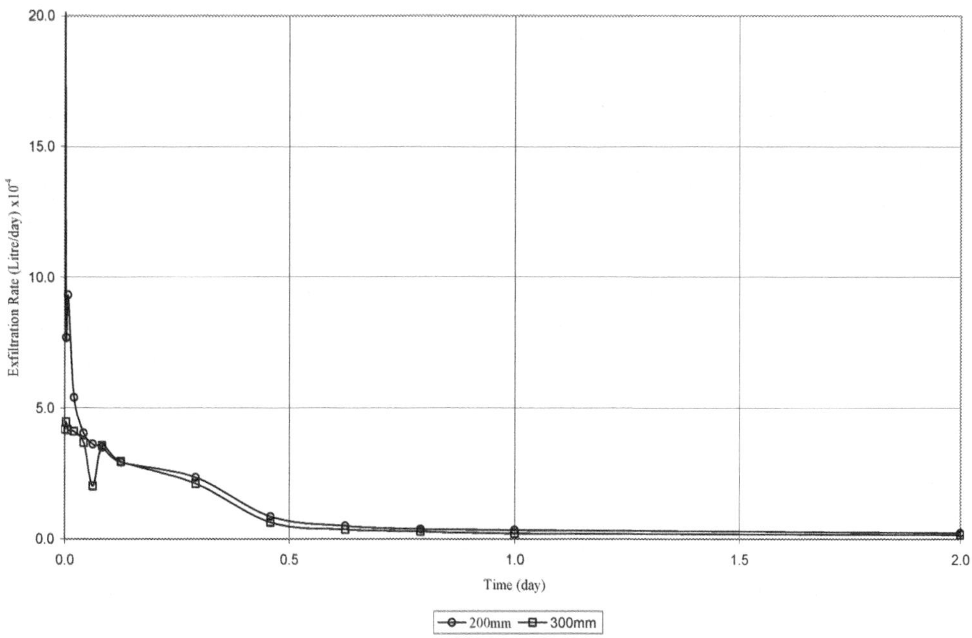

Figure 4 Exfiltration Rate Under Constant Leakage Area (20 cm^2) and varied soil bedding thickness

Figure 4 shows that the rates of exfiltration through the sand bedding were initially high for both set-up. This due to exfiltration taking place into clean sand. The exfiltration rate became almost constant within a day. The reduction in the rate of exfiltration for both experiments were similar. When comparing the results between 200 mm and 300 mm of soil bedding thickness, shows that the result of 200 mm thickness is slightly higher compared to the result of 300 mm thickness and sometimes the results are almost same. This indicates that the different thickness of the soil bedding did not affect the exfiltration rate.

Table 3 and Table 4 show the data for the exfiltration rate from leakage areas of 10 cm^2 and 30 cm^2 through a sand bedding thickness of 200 mm. Figure 5 shows the result of the rates of exfiltration from different leakage sizes but through a constant soil bedding thickness.

Table 3 Exfiltration Rate Under leak of 10 cm^2 and Soil Bedding Thickness of 200 mm at column reactor C

Time (min)	Exfiltration rate (Litre/day) x 10^{-4}							
	Test 1	Test 2	Test 3	Test 4	Test 5	**Average**	min	max
0	27.78	17.36	10.76	0.4861	0.8333	**11.44**	0.4861	27.78
5	14.24	15.97	11.81	0.3472	1.6667	**8.81**	0.3472	15.97
10	11.46	16.32	1.39	0.5556	1.6667	**6.28**	0.5556	16.32
30	7.67	15.63	1.88	0.1736	0.4167	**5.15**	0.1736	15.63
60	4.17	14.24	1.39	0.0694	0.2778	**4.03**	0.0694	14.24
90	2.99	13.19	1.18	0.0347	0.1736	**3.51**	0.0347	13.19
120	1.74	13.19	0.903	0.0347	0.0694	**3.19**	0.0347	13.19
180	1.18	9.86	0.694	0.2083	0.0347	**2.40**	0.0347	9.86
420	0.833	9.03	0.694	0.2083	0.0347	**2.16**	0.0347	9.03
660	0.521	2.57	0.486	0.0694	0	**0.73**	0	2.57
900	0.313	1.18	0.347	0.0347	0	**0.37**	0	1.18
1140	0.278	1.04	0.139	0.1389	0	**0.32**	0	1.04
1440	0.243	0.625	0.104	0.2778	0	**0.25**	0	0.625
2880	0.243	0.556	0.0694	0.0347	0	**0.18**	0	0.556

Table 4 Exfiltration Rate Under leak of 30 cm^2 and Soil Bedding Thickness of 200 mm at column reactor D

Time (min)	Exfiltration rate (Litre/day) x 10^{-4}							
	Test 1	Test 2	Test 3	Test 4	Test 5	**Average**	min	max
0	27.78	13.19	12.5	0.6944	0.5556	**10.94**	0.5556	27.78
5	17.88	20.83	8.33	1.0417	0.9028	**9.80**	0.9028	20.83
10	7.99	13.89	11.11	3.125	0.2778	**7.28**	0.2778	13.89
30	5.35	16.67	1.04	2.0833	0.0347	**5.04**	0.0347	16.67
60	2.71	14.58	1.32	0.9028	0.0347	**3.91**	0.0347	14.58
90	1.91	13.89	0.903	1.3194	0.0694	**3.62**	0.0694	13.89
120	1.11	14.58	0.833	2.0833	0.1389	**3.75**	0.1389	14.58
180	0.799	12.15	0.694	1.9444	0.2083	**3.16**	0.2083	12.15
420	0.625	10.42	0.486	2.1528	0.0139	**2.74**	0.0139	10.42
660	0.417	4.79	0.451	1.3889	0.0139	**1.41**	0.0139	4.79
900	0.278	2.5	0.313	0.9028	0	**0.80**	0	2.5
1140	0.278	2.08	0.278	0.4167	0	**0.61**	0	2.08
1440	0.139	1.25	0.104	0.0069	0	**0.30**	0	1.25
2880	0.174	1.04	0.0694	0	0	**0.26**	0	1.04

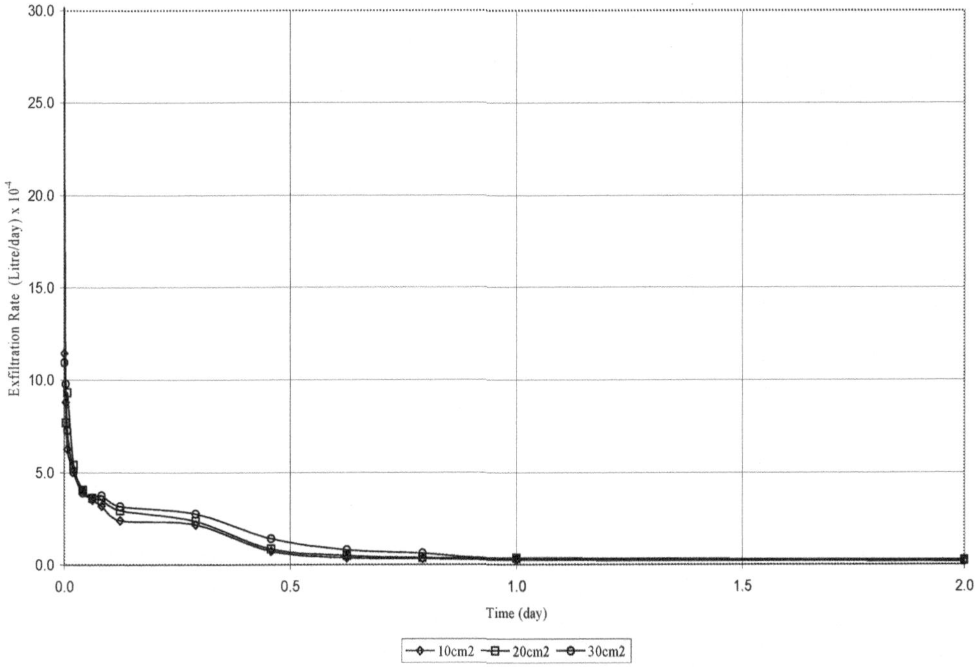

Figure 5 Exfiltration Rate Under Constant Soil Bedding (200 mm) and varied leakage areas

From the result in Figure 5 , the result of 10 cm^2 leakage area was slightly lower than the result of 20 cm^2 and 30 cm^2 leakage area and became almost same toward the end. It can be shown that the rate of exfiltration increases with increase in the size of the leakage but similar pattern was seen for the reduction in the rate of exfiltration regardless of the size of the leakage. In all the three cases the rate of exfiltration were initially high and became almost constant after one day.

Conlusions

From the results, the following conclusion can be drawn.

Firstly, it can be concluded that the thickness of the soil bedding will not affect the rate of exfiltration. In both cases, the accumulating of the biofilm at the bedding layer beneath the sewer leakage produced the clogging zone responsibled for the reduction in the exfiltration rate.

Secondly, the rate of exfiltration is governed by the size of the leakage area and the development of the biologically active clogging zone. From the experiments, it was found that the size of the leakage area was found to play an important role in determining the size of the clogging zone.

References

Davies, J.P., Clarke, B.A., Whiter, J.T. and Cunningham, R.J. (2001) Factors influencing the structural deterioration and collapse of rigid sewer pipes. *Urban Water*, **3**, 73-89.

Decker, J. and Risse, A. (1993) Investigations about qualitative and quantitative pollution load of subsoil, ground and surface water by leaking sewers. In Marsalek, J. and Torno, H. (Eds.) *Proc 6th Int. Conf. Urban Storm Drainage*, Sea Point Publishing, Victoria, British Columbia, Canada.

Eiswirth, M. and Hotzl, H. (1994) Groundwater contamination by leaking sewerage systems. *Water Down Under*, November 1994, Adelaide, Australia, 21-25.

Ellis, J.B. (2001) Sewer infiltration/exfiltration and interaction with sewer flows and receiving groundwater quality. In *Proc. IAHR/IWA 2nd Int. Conf. INTERURBA II, Interaction between sewers, treatment plants and receiving waters in urban areas*, Lisbon, 19-22 February, 311-319.

Ellis, J.B. and Revitt, M.D. (2001) Sewer losses and interactions with groundwater quality. *Wat. Sci. Tech.* **45** (3).

Ellis, J.B., Revitt, M.D., Lister, P. and Willgress, C. (2002) The effects of sediments on sewer exfiltration. In *Proc. of the 3rd Int. Conf. on sewer processes and networks*, Paris, 15-17 April, 147-154.

IWK(1998) Developing a Modern & Efficient Sewerage System for Malaysia. Kuala Lumpur: IndahWater Konsortium.

Vollertsen, J. and Hvitved-Jacobsen, T. (2002) Exfiltration from gravity sewers – a pilot scale study. In *Proc. of the 3rd Int. Conf. on sewer processes and networks*, Paris, 15-17 April, 155-162.

Part Two: Workshop Guidelines

Guidelines for abstracts

J.-L. Bertrand-Krajewski

Laboratoire URGC, INSA de Lyon, 34 avenue des Arts, 69621 Villeurbanne cedex, France.
email: jean-luc.bertrand-krajewski@insa-lyon.fr

Introduction

The following guidelines are given to help you to prepare abstracts submitted to conferences, workshops, etc. They should be considered as suggestions based on experience, and not as strict and mandatory rules.

A good abstract should give information about the following 7 elements:
- the title
- the authors
- the context
- the scientific questions
- the method and/or the approach
- the results and their discussion
- the conclusions and the perspectives.

According to the call for abstracts of the conference, other information may be requested: reference to conference topics or themes, keywords, literature citations, etc. It is obvious that you have to follow the prescriptions of the conference organisers.

The title

The title should be as informative as possible, but also as concise as possible: a title is always a compromise. But it is a critical compromise because it gives the reviewer his/her first feeling and ideas about the topic and the content of your abstract. Sometimes, a maximum length in words or lines is required by conference organisers. To save some words and to keep only valuable information in your title, avoid titles starting with words like: "Contribution to the study of...". According to the content of your abstract, fundamental words should appear in the title, like "...experiments in...", "...modelling of...", "...theory of...", etc.

The authors

You should give the names, affiliations and addresses of all authors. The corresponding author should appear very clearly, with his/her email address, fax and telephone numbers.

For many conferences, abstracts shall be kept anonymous for reviewers. Consequently, the title of the abstract, and names and addresses of authors shall be given on an additional or

 Sewer Networks and Processes within Urban Water Systems edited by J.-L. Bertrand-Krajewski, M. Almeida, J. Matos & S. Abdul-Talib. ISBN: 1 84339 506 1.

separate front page that will not be sent to the reviewers. Once again, follow the rules given by the conference organisers.

The context of the research

The context of the research means i) the field of your research, and ii) the existing or previous state of art or state of knowledge in this field. It should also contain some brief elements to explain why this field is important for scientific, technical, operational, ecological or economical reasons. In other words, what are the stakes of your work ?

Avoid two extreme cases:

i) do not start from scratch, like "it is well known since Greek and Roman civilisations that rainfall events occurring in cities may have consequences in terms of runoff and transport of substances....". You will need at least one page to reach the core of your work: remember that abstracts are very short, typically 1 or 2 pages.

ii) reciprocally, do not start with over-specialised elements that are not necessarily known by reviewers, like "Previous work in our institute has shown that the second-order derivative of the Smith-Parker-Wilkinson equation is absolutely critical for urban development...".

Topics or themes are usually given in calls for papers of conferences. If your abstract corresponds to one (or more) of these topics, you may use this information to start your abstract and to help the reviewer to identify and understand quickly your field and/or topics.

The scientific questions

In relation to the context, you should now explain what scientific question(s) you want to answer. Test of a hypothesis or of a theory, experimental or laboratory measurements to improve knowledge, to test or validate models, new techniques, new methods, new models, transposition of ideas, concepts, models, methods from another scientific field into your research field, etc.?

A clear definition of this(ese) question(s) is absolutely required.

The method and/or approach used

In this section, you should describe what methods, techniques, tools and approaches you use to provide answers to the above scientific questions. If necessary, some (brief) theoretical background elements may be useful for the reviewer.

Laboratory and/or field experiments ? What instruments do you use ? How do you proceed ? How do you process data, information, results ? What significant hypotheses or assumptions do you use ? What model or theory do you use ? What variables or parameters do you measure or analyse ? What kind of mathematical, analytical, or software tools do you use ? What kind of protocol do you apply ?

If your research needs successive steps or phases, explain how they are organised and their logic.

The results and their discussion

This section should provide a brief description of your results. You should give some data, tables, figures, synthetic results. If you have only preliminary or partial results, this should be explained. The results should then be discussed in order to show how they contribute to answer or not your above scientific questions, how they may validate or not your hypotheses, your model, etc., how they may generate new questions or problems.

The above paragraph implies that you have already obtained some results, which is not systematically the case. However, remember that abstracts which do not provide results, but only general ideas, concepts or plans have a very low probability to be selected by the reviewers. As the number of presentations is limited in conferences, you should always keep in mind the following question that the reviewer(s) shall answer: what is the potential interest for the audience to hear your presentation if, based on the abstract, there are no results but only intentions or plans ?

Conclusions and perspectives

This section is usually short in abstracts, but should provide some elements. What are the future tasks ? Is your research programme finished ? What are the new questions, the new problems, the possible extensions ?

It is of course not the aim of these guidelines to lead to fully formatted and "standardised" abstracts. However, some orders of magnitude could be given about the respective lengths of the last 5 sections, expressed hereafter in percents of the total length:

- the context	10 %
- the scientific questions	15 %
- the method and/or approach	25-40 %
- the results and their discussion	30-45 %
- the conclusions and the perspectives	5 %

Keep in mind that your abstract should be:

- informative: the reviewer has to answer the question "based on the abstract, what is the potential interest (in terms of results, knowledge, data, models, etc.) for the audience to listen to this presentation ?"

- attractive: beyond the scientific content of your abstract (which is, at least theoretically, objective), the reviewer has also to answer the question "based on the abstract, could one expect a good presentation or a good paper, i.e. a presentation or a paper that will be well structured, well organised, clear in terms of objectives, questions, results and discussion ?"

Guidelines for oral presentations and preparation of slides

J.-L. Bertrand-Krajewski

Laboratoire URGC, INSA de Lyon, 34 avenue des Arts, 69621 Villeurbanne cedex, France.
email : jean-luc.bertrand-krajewski@insa-lyon.fr

Introduction

The following guidelines are given to help you to prepare your oral presentations and your slides. They should be considered as suggestions based on experience, and not as strict and mandatory rules. As an oral presentation also reflects the personality of its author, it is obvious that rules should be adapted by each speaker. The following guidelines are provided to avoid lousy presentations : if you follow them, you increase your chance of giving an attractive and successful presentation, but it is not an absolute guarantee. Oral presentations remain a kind of art, and techniques are necessary but never sufficient...

Oral presentation

- you should speak distinctly and slowly.

- you should speak to the audience in front of you, and not to the laptop, to the overhead projector, or to the screen on the wall.

- if you want to show a detail on a slide, use a pointer or the mouse of your laptop, but do not turn to face the screen.

- you should vary the rythm and the intonation during your presentation. Monotonic presentations are absolutely boring for the audience : even if giving a scientific presentation is not a player performance, it should be living. However, keep always in mind that the scientific content remains the most important aspect : the most living presentation will never compensate poor scientific content or quality.

- when no microphone is available, you should speak loud enough to be heard from the last row in the room.

- if you do not feel comfortable to speak without notes (especially when you have to speak in a language which is not your mother-tongue), reading some notes is acceptable and better than a confusing and hesitating speech. However, if you read, you should avoid two major obstacles : i) reading too fast and ii) not looking at the audience. Nevertheless, if you speak about a topic you are familiar with (isn't it your work ?), do you really think that written notes are necessary ?

- you should keep an eye on your watch to respect your allocated time. If the chairperson asks you to conclude and if you know that you will be late, there is only one solution :

 Sewer Networks and Processes within Urban Water Systems edited by J.-L. Bertrand-Krajewski, M. Almeida, J. Matos & S. Abdul-Talib. ISBN: 1 84339 506 1.

remove some slides or overheads from your presentation and conclude! It is your responsibility to prepare a presentation that you can give within the allocated time.

- you should avoid remaining static during your presentation : some movements are welcome to have a living presentation. Do not be fixed to the mouse of your laptop! Inversely, do not be frenetic and agitated.

- as much as possible, depending on your degree of practice, check your corporal attitude : do not be aggressive or shy, keep open arms (hands in your pockets is not the best attitude), be relax and smile sometimes : it is always surprising to observe how the audience behaviour reflects your own attitude...

Slides

- the following rules are applicable for both traditional acetate overheads and PowerPoint slides.

- do not use vertical slides or overheads : there are always problems with screen format and dimensions!

- use simple fonts (like Arial...) that everybody can easily read from the last row in the room. Do not use fonts with thick and thin aspects (e.g. Times Roman font) : such fonts have been created to be read from short distance in newspapers and books, not for long distance in conference rooms! Legibility : this is the main objective.

- for slides titles, use CAPITAL (i.e. upper case) characters.

- minimum character size for slide titles : 30-32 pt.

- minimum character size for other words : 20-22 pt.

- maximum number of lines of text on each slide : 6-7 lines. If more ideas have to be introduced, use two slides or more.

- maximum number of words on each slide : 35-40 words. The slides shall not be used to display your full sentences but only some kind of keywords, fundamental elements or a thread which i) help the audience to follow your presentation, and ii) help you to remember what you have to say about each slide. If you need some more detailed written notes or sentences, use a separate sheet of paper or the "comments" tool in PowerPoint to read them : do not write them on your slides. Keep in mind that if too many words or sentences are written on your slides, the audience will try to read them and will not listen to you any longer!

- never use some simple "copy-paste" from previous written papers or documents, even for figures and tables. Adapt them specifically for your slides, by using appropriate fonts, character size, colours, thickness of lines and drawings, etc.

- use colours, but in an appropriate way : sufficient contrasts to facilitate legibility, not too many various colours which lead to confusion.

- as much as possible, avoid words and use graphs, figures, schemes, etc.

- on your first slide, some logos of institutions can be displayed if necessary. But it is sufficient to have them of this first slide : if you keep them on all slides, you stupidly loose space for the rest of the information. It is a scientific presentation, not an advertising show for trade marks!

- use the available surface as much as possible to display LARGE characters, LARGE graphs and LARGE figures. Keeping on all slides some logos or other background elements (e.g. like top, bottom or lateral banners) leads to a reduced usable space. One of the typical consequence of such a misuse of space is that you will use only 40 to 50 % of the surface and that you will frequently have to say something like : "I know that you can't read (or see) this figure (or this table), but ....".

- concerning tables of numerical values, the typical syndrome consists to copy-paste something like an Excel spreadsheet and to explain the audience that : "I know you can't read, but the only interesting values are at the intersection of column 7 and lines 5 and 6...". In other words, tables shall be SPECIFICALLY ADAPTED for slides : display ONLY the values about which you will give comments and explanations. Values that you do not use in your presentation shall not be on the slide. The paper in the proceedings is a very good solution to display more values...

- to check the legibility of your overheads, a simple test is suggested. Display your presentation on your computer screen and stand at a distance of 2.5 to 3 m from the screen : you should be able to read all texts, figures and tables without difficulty.

- new communication technologies are very nice and powerful, but sometimes also very uncertain : save and bring systematically TWO copies of your PowerPoint files. If you are either anxious or prudent or experienced (!), it could be interesting to have an "emergency" version of your slides printed on "old" acetate overheads.

- be careful if you plan to have some animations in your PowerPoint slides : check that the PowerPoint release that will be available during the conference is fully compatible with the release you used to prepare your slides. In case of doubt, it is better to remove complex animations because there is a high risk that they will not run properly.

- maximum number of slides : it depends on your own style of presentation, but typically should not be above approx. 1 slide per minute.

Guidelines for Session Chairpersons

J.-L. Bertrand-Krajewski

Laboratoire URGC, INSA de Lyon, 34 avenue des Arts, 69621 Villeurbanne cedex, France.
email: jean-luc.bertrand-krajewski@insa-lyon.fr

Introduction

This document aims to provide some basic information on the tasks a chairperson should ensure during a conference session. If the chairperson is also the rapporteur, he/she should read the companion document describing the tasks of a session rapporteur. However, as much as possible, it is recommended to have two persons, one for each position.

The main objective of the chairperson consists to manage the conference session in a smooth way and a kind atmosphere, by introducing the session and the speakers, by facilitating and managing answers from the audience and answers from the speakers, and by keeping the time schedule.

Before the session

Reading the papers

The chairperson should read the papers of the conference session he/she will chair. This will help him/her to introduce the session, to know its content and how some links and transitions may be suggested between the papers, and to prepare some questions for the authors.

Collecting information about speakers

As the chairperson will have to introduce the speakers during the session, it is recommended to collect some basic information about them: name and first name (always check with speakers how to pronounce properly their name and first name, especially for speakers who are coming from regions of the world the chairperson is not familiar with), institution, diploma and/or cursus, present and (if appropriate) former positions, mains research interests and activities, activities in national or international working groups, etc.

Checking presentation supports

Typically a half-day before the session (and at least 30 minutes for the first session), the chairperson should ask the speakers to check that everything is OK with their presentation: overheads and overhead projector, slides, PowerPoint files and computers, etc. Frequently, this checking is made with a specific person from the conference staff, but it is the responsibility of the chairperson to remember this to the speakers. In case of problems, he/she will have to re-organise the session.

 Sewer Networks and Processes within Urban Water Systems edited by J.-L. Bertrand-Krajewski, M. Almeida, J. Matos & S. Abdul-Talib. ISBN: 1 84339 506 1.

During the session

Checking the presence of all speakers

Typically 10 minutes before the session, the chairperson should be present in the conference room and check that all speakers are present with ready material and supports. This is the last moment to re-organise the session if necessary. It is also the moment to remember the speakers how much time they have for their oral presentations.

Introducing the session

When it is time to start the session, the chairperson should ask for silence if necessary, welcome the audience, declare the session opened, and give some words to present very briefly its topics and its main themes. The chairperson should also explain (if necessary) if questions will be asked after each presentation or grouped after the full set of presentations.

Introducing the presentations

For each presentation, the chairperson should introduce the speaker (with the information collected previously) and his/her presentation. During this short introduction (1 minute maximum), the first speaker of the session should open his/her PowerPoint file or prepare his/her overheads.

Watching at time

The chairperson is in charge of keeping the time schedule. If some speakers speak too long, the last ones will have less time, or the sessions will not end on due time, or it will lead to problems if participants want to listen to presentations in parallel sessions. Keeping time is thus a very critical aspect.

When approx. only 2 minutes remain for the speaker, the chairperson should inform him/her about that (by oral announcement, or by showing a sheet of paper, or by any other appropriate means). And when allocated time is elapsed, the chairperson has, if necessary, to ask the speaker to conclude. The chairperson should do this kindly, but firmly. In case the speaker is not able to conclude after 1 or 2 additional minutes, the chairperson has to stop him/her.

Discussion

After the speaker has finished his/her presentation, the chairperson should open the discussion phase and ask if some participants have questions or comments about the presentation. If no questions are asked during the next seconds, the chairperson may start himself/herself: this is the reason why it is important for him/her to have prepared some questions in advance after having read the papers.

The chairperson should look carefully at the audience in order to look at participants who want to ask questions, and to give them their turn to ask questions. Under some circumstances, the chairperson should re-formulate some questions or help for mutual understanding when participants and speakers are facing to translation problems.

The chairperson should ask the participants to introduce themselves briefly (name, institution, city and/or country) before they ask their questions.

If a microphone is necessary to ask questions, the chairperson should wait until the participant has access to the microphone before he/she asks his/her question, so that all the audience can understand it clearly.

The chairperson should keep questions and answer under control. Questions or comments should be relatively short in time, and answers from the speakers too.

The chairperson should also stop questions on due time, to keep the time schedule for the next speakers or the next sessions. If many participants want to ask questions and if sufficient time is not available, the chairperson should decide when the last question will be asked and should announce this to the audience: "Sorry, this will be the last question", before to give his/her turn to the participant authorised to ask the last question. If many questions are pending, it is kind to suggest that participants may meet the speaker after the session. At the end of the questions, he should thank the speaker and start to introduce the next one.

During the last question, the next speaker of the session should open his/her PowerPoint file or prepare his/her overheads.

At the end of the session

Closure of the session

After the last answer to the last question, the chairperson should close the session. A traditional way to do this consists to thank all speakers for their presentations and all participants for their questions and comments.

Afterwards, and depending on the context, the chairperson should:

- announce the duration and time of the end of the coffee break or of the lunch
- announce the time at which the next session will start
- make some announcements as asked by conference organisers, other chairpersons, etc.

Guidelines for Session Rapporteurs

J.-L. Bertrand-Krajewski

Laboratoire URGC, INSA de Lyon, 34 avenue des Arts, 69621 Villeurbanne cedex, France.
email: jean-luc.bertrand-krajewski@insa-lyon.fr

Introduction

This document aims to provide some basic information on the tasks a rapporteur should ensure during a conference session. If the rapporteur is also the chairperson, he/she should read the companion document describing the tasks of a session chairperson. However, as much as possible, it is recommended to have two persons, one for each position.

The objective of the rapporteur consists to report the main concepts and results given by the speakers, and to keep track of the questions and answers of the session. At the end of the session, he/she has to prepare a short synthesis report (typically 1 page) with the most significant conclusions of the session, in terms of ideas, concepts, results, scientific questions and problems, debates, etc.

Before the session

Reading the papers

The rapporteur should read the papers of the conference session he/she will chair. This will help him/her to get familiarity with the topic and with the content of the presentations, to pre-identify the most interesting information and some links between the papers, and to prepare his/her final report.

During the session

During the presentations

During each presentation, the rapporteur should extract the most significant information, and should look carefully at information which is given during the presentation but which was not written in the paper. This occurs frequently, as new results have been obtained by the authors between the time they wrote the paper for the conference proceedings and the time of the presentation. As much as necessary, it could be useful for the rapporteur to have a copy of the PowerPoint files used for the presentations.

During the discussion

During the discussion, the rapporteur should keep track of all questions, answers and debates.

After the session

Preparation of the report

After the session, typically during the next 24 hours to keep a fresh memory of the session and to avoid to postpone the work, the rapporteur should prepare a *synthesis report* which includes the title of the session, the name of the chairperson and the name of the rapporteur, *key ideas, significant results or useful information for all presentations*. This synthesis should be based on the papers, on the presentations and on the track of the questions and answers written by the rapporteur, but should focus only on the main elements, and *not report all details of the above elements*. The ability to prepare a synthetic report is one aspect of the art of reporting. Instead of long sentences, it is possible to provide bullet lists. For example some lists with up to 5-6 items in each list:

- a list of the main initial scientific questions and problems.
- a list of most significant new data and results, new knowledge, new models, new contributions, etc.
- a list of remaining and/or new scientific questions.
- a list of possible interactions, doubts, conflicts between data, results, interpretation, etc.
- etc.

The rapporteur may add some personal conclusions, but they shall clearly appear as such.

It is frequently very valuable to ask the session chairperson to review and to complete the report.

Guidelines for "Advocatus Diaboli"

Maria do Céu ALMEIDA*, Jean-Luc BERTRAND-KRAJEWSKI**

* LNEC, DH-NES, 101 avenida do Brasil, 1700-0166 LISBOA, Portugal, email:mcalmeida@lnec.pt

** Laboratoire URGC, INSA de Lyon, 34 avenue des Arts, 69621 Villeurbanne cedex, France. email: jean-luc.bertrand-krajewski@insa-lyon.fr

Introduction

This document aims to provide some basic information on the tasks an *"advocatus diaboli"* should ensure during a conference session. If the *"advocatus diaboli"* is also the chairperson, he/she should read the companion document describing the tasks of a session chairperson. However, as much as possible, it is recommended to have two persons, one for each position.

The objective of an *"advocatus diaboli"* consists in triggering the discussion after a presentation or after a session. Experience shows that this function often is minor simply because discussion usually kicks off without delays. If not, this person should be prepared to put forward some relevant comments or questions. However, the idea is not to dominate the discussion but raise stimulating questions, eventually a little provocative, within the scope of the presentation(s).

Before the session

Reading the papers

The *"advocatus diaboli"* should read the papers of the conference session where he/she is assigned that function. This will help him/her to get familiarity with the topic and with the content of the presentations, identify interesting comments or questions that can be used if discussion does not go on naturally.

During the session

During the presentations

During each presentation, the *"advocatus diaboli"* should look carefully at information which is given during the presentation but which was not written in the paper to eventually alter/complement the comments or questions previously prepared. This occurs frequently, as new results have been obtained by the authors between the time they wrote the paper for the conference proceedings and the time of the presentation. It could be useful for the *"advocatus diaboli"* to have a copy of the PowerPoint files used for the presentations.

During the discussion

During the discussion, the *"advocatus diaboli"* should wait to see if questions arise and, if not, start the discussion. Other questions can be put forward, participating as any other participant in the discussion within time limits.

The "*advocatus diaboli*" should ask questions or give comments which should help the speaker and/or the audience:

- to clarify and explain any point that remains unclear after the presentation

- to extract as much information as possible from the facts, experience, results, etc. given by the author/speaker, and to question their validity, representativity, reproducibility, transferability, etc.

- to point some critical hypotheses, assumptions, theories or viewpoints, and protocols used by the author/speaker that may significantly affect scientific work and scientific conclusions. This is particularly important as, frequently, options or choices are not seriously discussed in conferences. Many ad-hoc options, choices, etc. are made without being rigorously discussed. In other words, the objective is to deepen the scientific methodology, and to make clear how scientific work and thinking is made. It may eventually refer to epistemology. This is particularly valuable for modelling.

- to refer to existing or previous work on the same topics, especially if different conclusions or approaches are/have been used. Reference to other national and international research is frequently a weak point, for many (but bad…) reasons: too many papers, conferences and journals, too much grey literature, not enough time, etc. This is consequently important that the "*advocatus diaboli*" contributes to a fruitful literature and cross-reference process.

After the session

Contribution to the session report

It might be appropriate that the *"advocatus diaboli"* contributes to review the session report prepared by the rapporteur.

Keyword Index

IWA Publishing's authorised EU representative for General Product Safety Regulations is Diane D'Arras, 15 rue Duret, 75116 Paris, France, e-mail: safety@iwap.co.uk.

Printed and bound by CPI Group (UK) Ltd, Croydon, CR0 4YY
12/05/2026
02108880-0016